Basic Civil and Environmental Engineering

Basic Civil and Environmental Engineering

Ramesh Shah

RANDOM

RANDOM PUBLICATIONS

NEW DELHI (INDIA)

Basic Civil and Environmental Engineering

ISBN 978-93-5111-428-4

Published in 2014 in India by

RANDOM PUBLICATIONS

4376-A/4B, Gali Murari Lal, Ansari Road
New Delhi-110 002
Phone : +91-11-43580356, +91-11-23289044
e-mail: randomexports@gmail.com, sales@randompublications.com, info@randompublications.com

Reprinted 2022

Type Setting by: Friends Media, Delhi-110089
Digitally Printed at : Replika Press Pvt. Ltd.

Preface

Civil engineering is a professional engineering discipline that deals with the design, construction, and maintenance of the physical and naturally built environment, including works like roads, bridges, canals, dams, and buildings. Civil engineering is the oldest engineering discipline after military engineering, and it was defined to distinguish non-military engineering from military engineering. It is traditionally broken into several sub-disciplines including environmental engineering, geotechnical engineering, geophysics, geodesy, control engineering, structural engineering, transportation engineering, earth science, atmospheric sciences, forensic engineering, municipal or urban engineering, water resources engineering, materials engineering, offshore engineering, quantity surveying, coastal engineering, surveying, and construction engineering. Civil engineering takes place on all levels: in the public sector from municipal through to national governments, and in the private sector from individual homeowners through to international companies. Engineering has been an aspect of life since the beginnings of human existence. The earliest practice of civil engineering may have commenced between 4000 and 2000 BC in Ancient Egypt and Mesopotamia (Ancient Iraq) when humans started to abandon a nomadic existence, creating a need for the construction of shelter. During this time, transportation became increasingly important leading to the development of the wheel and sailing.

Throughout ancient and medieval history most architectural design and construction was carried out by artisans, such as stonemasons and carpenters, rising to the role of master builder. Knowledge was retained in guilds and seldom supplanted by advances. Structures, roads and infrastructure that existed were repetitive, and increases in scale were incremental. Civil engineers typically possess an academic degree with a major in civil engineering. The length of study for such a degree is usually three to five years and the completed degree is usually designated as a Bachelor of Engineering, though some universities designate the degree as a Bachelor of Science. The degree generally includes units covering physics, mathematics, project management, design and specific topics in civil engineering. Initially such topics cover most, if not all, of the sub-disciplines of civil engineering. Students then choose to specialize in one or more sub-disciplines towards the end of the degree. While an Undergraduate (BEng/BSc) Degree will normally provide successful students with industry accredited qualification, some universities offer postgraduate engineering awards (MEng/MSc) which allow students to further specialize in their particular area of interest within engineering.

Civil engineers design and build many of the things that we use every day. Civil engineers work in the structures, transport, energy and water storage design works. Civil engineers need to be able to be good at maths, be able to communicate their ideas to people and be able to design. They need to be able to think in 3D.

I would like to thank my team for standing beside me throughout my career and writing this book. My special thanks go to "Random Publications" who have published the book.

—Ramesh Shah

Contents

1

Design Options

The design options for welded and bolted connections may be configured in the Options menu (Joints > Options). This dialogue box can be accessed via the following menu options:

- In CYPECAD and its Integrated 3D structures:
    - Beam Definition tab > Job > Joint analysis options
    - Results tab > Joints > Joint analysis options
- In Metal 3D:
    - Joints > Options

This dialogue box contains three tabs:

- ***Non-prestressed Bolts:*** Contains the design options for non-prestressed bolts (Bolt series code, Bolt steel class and Available diameters). Options selected here will only affect bolted connection design.
- ***Prestressed Bolts:*** Contains the design options for prestressed bolts (Bolt series code, Bolt steel class, Available diameters and Friction surface class). Additionally, the type of friction surface must be indicated to check the bolt for slipping. Within this dialogue, the program also defines the properties of the selected surface in accordance with the article of the selected code. Options selected here will only affect bolted connection design.
- ***Stiffeners:*** Contains two option groups which configure the layout of the stiffeners:
    - ***Stiffeners at Haunch Ends:*** Contains two options which when activated force the program to always provide

stiffeners, one in column to beam connections and the other in ridge connections. Affects welded connections designed with the Joints I module and bolted connections designed with the Joints II module.

Regardless of whether this option is activated or not, the program always provides stiffeners at the ends of the haunches if the design requires their presence.

- ***Stiffeners for Beams Fixed to the Column Web:*** Activates the trimming of stiffeners for beams fixed to column webs. It only affects welded connections designed with the Joints III module and bolted connections using the Joints IV module (the other modules do not contemplate connections consisting of beams fixed to column webs).

  This option does not imply changes in the structural check but only takes into account aspects related to the aesthetics and with the number of trimming operations undertaken in the fabrication process. Upon activating the trimming option, it will be carried out if the following conditions are met:

  — The smallest side of the trim has to be greater than 10mm.

  — The angle formed between the inclined side of the trim and the plane perpendicular to the web has to be greater than 15 degrees.

Prestressed and Non-prestressed Bolted Connections

CYPECAD, Metal 3D and Integrated 3D structures of CYPECAD allow bolted connections to be designed using prestressed or non-prestressed bolts. This option is located in the Analysis menu (Analysis > Analyse) or in the joints design dialogue box (Joints > Analyse).

When the structure is analysed taking into account seismic effects, the program either recommends or imposes the use of prestressed bolts depending on the selected seismic code. If the code states the use of prestressed bolts is obligatory, the program selects this option and does not allow for it to be deactivated. If the seismic code does not state it is compulsory to use prestressed bolts, the program reminds the user it is advisable to use these as is recommended by Eurocode 8.

When bolted connections are used, the program also indicates the friction surface class to be used in case that selected by the user in the design options is not adequate.

If the warnings are not taken into account by the user, it is possible for the prestressed bolted connection to not be designed by the program. In this case, when the user consults the joints after the analysis, the program indicates as errors the reasons why the design has not been carried out.

Bolted Connection Design

General Properties

If during the analysis process of the structure, nodes are detected whose configuration is adjusted to those types resolved by the program, they will be designed and detail drawings, 3D views, code check reports and take-off of these designs will be provided.

The program designs the dimensions of the plates and stiffeners, the optimum number and layout of the bolts and the throat thicknesses of the welds required for the correct stress transmission in the connection. The program designs the bolted connections using a front plate or a lateral plate:

- Moment connections using front plate. This type of solution is used by the program for fixed connections between beams and columns, for element splices and ridge beam connections.
- Simple connections using lateral plates. This type of solution is used by the program for pinned connections between beams and column webs or flanges, and webs of other beams (secondary beams).
- Simple connections using front plates. This type of solution is used by the program for beams supported at column ends forming a pinned connection.

Non-deformable Bar Ends

When welded or bolted connections are designed, or when the structure is analysed together with the joints, the program carries out, at each node of the job, the spatial layout of the bars and its fixity conditions in such a way that, for steel bars, the dimensions of the nodes are established and generates the rigid ends in which the portions of the bars are considered to not deform due to them being contained within the node.

Rotational stiffnesses at element ends: The program allows for fixity coefficients xy and xz or rotational stiffnesses to be assigned to bar ends (bars or set of aligned bars forming one element). Defining the rotational stiffness allows for the joints to be modelled and whereby their stiffness to rotate is fundamental to bear into account, as is in the case of bolted connections.

For each bolted connection that is designed, the program analyses (for all the acting force combinations) the rotational stiffness of the end of each fixed element and suggests a value with which the structure is to be reanalysed. The proposed stiffness will be the smallest amongst those calculated at each element end, which corresponds to the one with the greatest positive or negative moment.

After the analysis and for elements fixed to the bolted connections, if the user has not defined the rotational stiffness or if the introduced value differs in more than 20% from that proposed by the program, a warning is emitted. This is also displayed as a warning in the report at the end of the analysis or in the option Show error messages in the Analysis menu.

Using the Rotational stiffnesses option in the Joints menu, the user can manage and substitute the values that were introduced instead of those proposed by the program (at the ends of elements with moment bolted connections provided by the program). This option activates a dialogue box with two possibilities:

Check the Assigned Rotational Stiffness Values

Displays the element ends whose rotational stiffnesses are adjusted to those proposed by the program in green; whilst those with a stiffness differing in more than 20% to that proposed by the program are shown in red.

By placing the cursor over one of these element ends, the program displays a tag with the messages associated to that end and highlights in blue all the element ends of the structure that follow the same behaviour, moment-rotation diagram and assigned rotational stiffness.

By selecting the end of the desired element, a dialogue box is displayed containing information on the type of fixity assigned to the end of the element and the rotational stiffness value proposed by the program, as well as the behaviour graph of the connection expressed by a moment-rotation curve.

In this dialogue box, the original values can be maintained or modified. If they are modified, all the elements of the same type will be affected (those highlighted in blue when an end was selected). This dialogue box also contains a help button (icon) in which the criteria followed by the program to obtain the rotational stiffness proposed by the user is thoroughly explained.

- *Assign the Proposed Rotational Stiffness Value to All the Elements:* Automatically assigns the rotational stiffness values proposed by the program to all the following element ends (only at element ends fixed with bolted connections).
  - o Those whose introduced rotational stiffness differs in more than 20% to that proposed.
  - o Those for which the user has not indicated any rotational stiffness.

The program then allows to consult the element ends in the same way as is done using the Check the assigned rotational stiffness values option.

After assigning the rotational stiffnesses proposed by the program to the desired element ends, the structure must be reanalysed so the new force distribution due to the change in the rotational stiffnesses can be taken into account. The program emits a warning of this situation.

Design Criteria for Prestressed Bolted Connections

The shear check in the transverse section of the bolts is substituted for an Ultimate Limit State (U.L.S.) slip check in the case of prestressed bolted connections and the tension and slip interaction is analysed.

When the user activates seismic action for the analysis of the structure, the fixed prestressed bolted connections are designed for Failure Mode 1, i.e., failure will occur due to plastic failure of the front plate of the connection and not due to bolt failure. Therefore, when the seismic action is taken into account, bolts are designed using greater diameters to guarantee enough rotation capacity.

Joint Consultation

Following the analysis, the joints that have been designed by the program can be consulted.

Metal 3D and CYPECAD place different coloured circles at the nodes to indicate whether or not all the connections of the node have been designed, if there are only a few that have been designed or if the node does not contain any designed connections. If the cursor is brought close to a node in which there are connections that have been designed, an information window will appear indicating the types of connections designed for that node. By clicking on the node a dialogue box with three tabs is displayed containing the following information:

- The construction details of the resolved connections
- A report of the checks and take-off of the resolved connections
- Real 3D views of the connections. It is possible to visualise a 3D view of each connection designed by the program with an isometric or conical perspective. The elements making up the joint (columns, beams, stiffeners, welds, bolts and plates) are drawn in different colours. Prestressed bolts and non-prestressed bolts are also represented in two colours to differentiate between them. The user may also freely rotate and amplify the 3D view. These characteristics greatly help the user to understand the assembly of the joint. The 3D view of the joints can be visualised by selecting the 3D view tab that appears in the bottom part of the window that is activated upon consulting the joint.

If the cursor is brought close to a node in which there are no designed connections, but belongs to one of the connections recognised by the program, a window appears informing of the causes that have prevented that joint from being designed.

Baseplates designed using the Joints II module

The Joints II. Bolted. Warehouses with rolled and welded steel I sections designs welded baseplates of columns on pad footings, pile caps and slabs.

The Joints II module also designs the baseplate connection points of integrated 3D structures, if these are located on footings, pile caps, slabs, shear walls, walls, concrete columns or beams (in the case of these last four elements, they are designed if no other steel beams or steel bars of the integrated 3D structures connect at that node).

The properties of the designed baseplates include:

- *Baseplate Types:* Centred baseplates for rolled steel and welded I sections are designed.
- *Welds:* Includes the analysis and design of the welds between the plate, stiffeners, column and bolts.
- *Automatic Match:* Automatically matches all the baseplates of a job (bearing in mind the section type, forces and external fixities). This way, and without the user having to intervene, the number of different types of baseplates is reduced, hence obtaining more uniform results.
- *3D View with Highlighted Elements and Welds:* It is possible to obtain a 3D view on screen whereby the baseplate, column,

stiffeners, bolts, factory welds and insitu welds are displayed in different colours, in a similar manner as to how joints are represented between I sections. This has been incorporated to help the user understand how the support is to be assembled.

- *Baseplate Layout:* A layout diagram of the baseplate is generated in which the details of the designed welds and stiffeners are displayed. This layout diagram can be included in the job drawings.
- *Takeoff and Code Check Reports:* Takeoff and code check reports are generated of baseplates that have been resolved. These are integrated with the rest of joints that have been designed.

Bolted Joints Report

Metal 3D and CYPECAD generate a joints report with the following data:

- Welded connection specifications
    - Code
    - Materials
    - Construction layout
    - Checks
- Bolted connection specifications
    - Code
    - Materials
    - Construction layout (includes conditions and possible fastening methods of prestressed bolts)
    - Checks
- References and symbols
- Baseplate checks
- Relationship of the designed joints
- Calculations
    - Construction detail of each type of joint
    - Description of the components of each type of joint
    - Check results of each type of connection
    - Weld, plate and bolt take-off for each connection type
- Complete take-off of welds, plates and bolts of the designed joints

Bolted Joints Drawings: The construction details of the joints designed by the program can form part of the drawings of the structure. The Joints drawings include the following elements:

- Construction detail of the joint
- Schedule with specifications of the welds in steel structures
    - o Code
    - o Materials
    - o Construction layout
    - o Fastening procedure of prestressed bolts
    - o Checks
- Reference and symbols table
- Complete take-off of welds, plates and bolts of the designed joints

Welded Connections

Weldability: The quality of a weld is dependent on the combination of materials used for the base material and the filler material. Not all metals are suitable for welding, and not all filler metals work well with acceptable base materials.

Steels

The weldability of steels is inversely proportional to a property known as the hardenability of the steel, which measures the ease of forming martensite during heat treatment. The hardenability of steel depends on its chemical composition, with greater quantities of carbon and other alloying elements resulting in a higher hardenability and thus a lower weldability. In order to be able to judge alloys made up of many distinct materials, a measure known as the equivalent carbon content is used to compare the relative weldabilities of different alloys by comparing their properties to a plain carbon steel. The effect on weldability of elements like chromium and vanadium, while not as great as carbon, is more significant than that of copper and nickel, for example. As the equivalent carbon content rises, the weldability of the alloy decreases. The disadvantage to using plain carbon and low-alloy steels is their lower strength there is a trade-off between material strength and weldability. High strength, low-alloy steels were developed especially for welding applications during the 1970s, and these generally easy to weld materials have good strength, making them ideal for many welding applications.

Stainless steels, because of their high chromium content, tend to behave differently with respect to weldability than other steels. Austenitic grades of stainless steels tend to be the most weldable, but they are especially susceptible to distortion due to their high coefficient of thermal expansion. Some alloys of this type are prone to cracking and reduced corrosion resistance as well. Hot cracking is possible if the amount of ferrite in the weld is not controlled to alleviate the problem, an electrode is used that deposits a weld metal containing a small amount of ferrite. Other types of stainless steels, such as ferritic and martensitic stainless steels, are not as easily welded, and must often be preheated and welded with special electrodes.

Arc Welding

The oldest of the Welding processes yet Arc Welding is still a very popular welding method. The Arc Welder, either Transformer or Inverter Welder is simple to set up and operate.

Arc welding processes use a welding power supply to create an electric arc between an electrode and the base material to melt metals at the welding point. They can use either direct (DC) or alternating (AC) current, and consumable or non-consumable electrodes. The welding region is sometimes protected by some type of inert or semi-inert gas, known as a shielding gas, and filler material is sometimes used as well.

One of the most common types of arc welding is shielded metal arc welding (SMAW), which is also known as manual metal arc welding (MMA) or stick welding. Electric current is used to strike an arc between the base material and consumable electrode rod, which is made of steel and is covered with a flux that protects the weld area from oxidation and contamination by producing CO_2 gas during the welding process. The electrode core itself acts as filler material, making a separate filler unnecessary. The process is very versatile, requiring little operator training and inexpensive equipment. However, weld times are rather slow, since the consumable electrodes must be frequently replaced and because slag, the residue from the flux, must be chipped away after welding. Furthermore, the process is generally limited to welding ferrous materials, though speciality electrodes have made possible the welding of cast iron, nickel, aluminium, copper, and other metals. The versatility of the method makes it popular in a number of applications, including repair work and construction.

Gas Metal arc Welding (GMAW), also Known as Metal Inert Gas (MIG) Welding

The fastest of the Welding processes; Mig Welding is therefore the most widely used. The Mig Welder is simple to set up and Welding is straightforward. This is a semi-automatic or automatic welding process that uses a continuous wire feed as an electrode and an inert or semi-inert gas mixture to protect the weld from contamination. Since the electrode is continuous, welding speeds are greater for GMAW than for SMAW. However, because of the additional equipment, the process is less portable and versatile, but still useful for industrial applications. The process can be applied to a wide variety of metals, both ferrous and non-ferrous. A related process, flux-cored arc welding (FCAW), uses similar equipment but uses wire consisting of a steel electrode surrounding a powder fill material. This cored wire is more expensive than the standard solid wire and can generate fumes and/ or slag, but it permits higher welding speed and greater metal penetration.

Gas Tungsten arc Welding (GTAW), or Tungsten Inert Gas (TIG) Welding

High quality is the key advantage in Tig Welding. Recent drops in Tig Welder prices have now brought this excellent Welding process within reach of all. This is a manual welding process that uses a non-consumable electrode made of tungsten, an inert or semi-inert gas mixture, and a separate filler material. Especially useful for welding thin materials, this method is characterised by a stable arc and high quality welds, but it requires significant operator skill and can only be accomplished at relatively low speeds. It can be used on nearly all weldable metals, though it is most often applied to stainless steel and light metals. It is often used when quality welds are extremely important, such as in bicycle, aircraft and naval applications. A related process, plasma arc welding, also uses a tungsten electrode but uses plasma gas to make the arc. The arc is more concentrated than the GTAW arc, making transverse control more critical and thus generally restricting the technique to a mechanized process. Because of its stable current, the method can be used on a wider range of material thicknesses than can the GTAW process, and furthermore, it is much faster. It can be applied to all of the same materials as GTAW except magnesium, and automated welding of stainless steel is one important application of the process. A variation of the process is plasma cutting, an efficient steel cutting process.

Submerged Arc Welding (SAW)

Is a high-productivity welding method in which the arc is struck beneath a covering layer of flux. This increases arc quality, since contaminants in the atmosphere are blocked by the flux. The slag that forms on the weld generally comes off by itself, and combined with the use of a continuous wire feed, the weld deposition rate is high. Working conditions are much improved over other arc welding processes, since the flux hides the arc and no smoke is produced. The process is commonly used in industry, especially for large products. Other arc welding processes include atomic hydrogen welding, carbon arc welding, electroslag welding, electrogas welding, and stud arc welding.

Plasma Cutting

A Plasma Cutter will slice most metals. A Plasma Cutter is the ideal tool for cutting metal prior to Welding.

Safety Issues

Welding, without the proper precautions, can be a dangerous and unhealthy practice. However, with the use of new technology and proper protection, the risks of injury and death associated with welding can be greatly reduced. Because many common welding procedures involve an open electric arc or flame, the risk of burns is significant. To prevent them, welders wear protective clothing in the form of heavy leather gloves and protective long sleeve jackets to avoid exposure to extreme heat and flames. Additionally, the brightness of the weld area leads to a condition called arc eye in which ultraviolet light causes the inflammation of the cornea and can burn the retinas of the eyes. Goggles and helmets with dark face plates are worn to prevent this exposure, and in recent years, new helmet models have been produced that feature a face plate that self-darkens upon exposure to high amounts of UV light. To protect bystanders, transparent welding curtains often surround the welding area. These curtains, made of a polyvinyl chloride plastic film, shield nearby workers from exposure to the UV light from the electric arc, but should not be used to replace the filter glass used in helmets.

Welders are also often exposed to dangerous gases and particulate matter. Processes like flux-cored arc welding and shielded metal arc welding produce smoke containing particles of various types of oxides. The size of the particles in question tends to influence the toxicity of the fumes, with smaller particles presenting a greater danger.

Additionally, many processes produce various gases, most commonly carbon dioxide and ozone, and fumes that can prove dangerous if ventilation is inadequate. Furthermore, because the use of compressed gases and flames in many welding processes pose an explosion and fire risk, some common precautions include limiting the amount of oxygen in the air and keeping combustible materials away from the workplace.

2

Design of Welds

Design of Butt Welds

For butt welds the most critical form of loading is tension applied in the transverse direction. It has been observed from tests conducted on tensile coupons containing a full penetration butt weld normal to the applied load that the welded joint had higher strength than the parent metal itself. The yield stress of the weld metal and the parent metal in the HAZ region was found to be much higher than the parent metal.

The butt weld is normally designed for direct tension or compression. However, a provision is made to protect it from shear. Design strength value is often taken the same as the parent metal strength. For design purposes, the effective area of the butt-welded connection is taken as the effective length of the weld times the throat size. Effective length of the butt weld is taken as the length of the continuous full size weld. The throat size is specified by the effective throat thickness.

For a full penetration butt weld, the throat dimension is usually assumed as the thickness of the thinner part of the connection. Even though a butt weld may be reinforced on both sides to ensure full cross-sectional areas, its effect is neglected while estimating the throat imensions. Such reinforcements often have a negative effect, producing stress concentration, especially under cyclic loads.

Unsealed butt welds of V, U, J and bevel types and incomplete penetration butt welds should not be used for highly stressed joints and joints subjected to dynamic and alternating loads. Intermittent butt welds are used to resist shear only and the effective length should not be less than four times the longitudinal space between the effective length of welds nor more than 16 times the thinner part. They are not to be used in locations subjected to dynamic or alternating stresses. Some modern codes do not allow intermittent welds in bridge structures. For butt welding parts with unequal cross sections, say unequal width, or thickness, the dimensions of the wider or thicker part should be reduced at the butt joint to those of the smaller part. This is applicable in cases where the difference in thickness exceeds 25 % of the thickness of the thinner part or 3.0 mm, whichever is greater. The slope provided at the joint for the thicker part should not be steeper than one in five. In instances, where this is not practicable, the weld metal is built up at the junction equal to a thickness which is at least 25 % greater than the thinner part or equal to the dimension of the thicker part. Where reduction of the wider part is not possible, the ends of the weld shall be returned to ensure full throat thickness. Stresses for butt welds are assumed same as for the parent metal with a thickness equal to the throat thickness (Cl.10.5.7.1). For field welds, the permissible stresses in shear and tension calculated using a partial factor ^{3}mw of 1.5. (Cl.10.5.7.2)

Design of Fillet Welds

Fillet welds are broadly classified into side fillets and end fillets. When a connection with end fillet is loaded in tension, the weld develops high strength and the stress developed in the weld is equal to the value of the weld metal. But the ductility is minimal. On the other hand, when a specimen with side weld is loaded, the load axis is parallel to the weld axis. The weld is subjected to shear and the weld shear strength is limited to just about half the weld metal tensile strength. But ductility is considerably improved. For intermediate weld positions, the value of strength and ductility show intermediate values.

Design of Tension Members

Tension Member

Tension members are structural elements that are subjected to axial tensile forces. They are usually used in different types of structures. Examples of tension members are: bracing for buildings and bridges, truss members, and cables in suspended roof systems.

In an axially loaded tension member, the stress is given by:

F = P/A

where P is the magnitude 10 (Hati) of the load and A is the cross-sectional area. The stress given by this equation is exact, knowing that the cross section is not adjacent to the point of application of the load nor having holes for bolts or other discontinuity. For example,

if you are given a 8 x 11.5 plate is used as a tension member (section a-a) and it is connected to a gusset plate with two 7/8-inch-diameter bolts (section b-b). So,

The area at section a - a (gross area of the member) is 8 x ½ = 4 in2

However, the area at section b - b (net area) is (8 – 2 x 7/8) x ½ = 3.12 in2 knowing that the higher stress is located at section b - b due to its smaller area.

To understand more about tension member, it would be useful to look at the stress-strain behaviour of steel:

Where, F_y is the yield stress, F_{su} is the ultimate stress, ε_y is the yield strain and ε_{su} is the ultimate strain.

Designing a Tension Member

In order to design tension members, it is important to analyse how the member would fail under both yielding (excessive deformation) and fracture which considered the limit states. The limit state that produces the smallest design strength is considered the controlling limit state; which also prevent the steel structure from failure.

Using AISC (American Institute of Steel Construction), we could obtain the recommended load and resistance factor design approaches.

The ultimate load on a structure can be calculated from one of the following combination:

$1.4\ D$

$1.2\ D + 1.6\ L + 0.5\ (L_r \text{ or } S)$

$1.2\ D + 1.6\ (L_r \text{ or } S) + (0.5\ L \text{ or } 0.8\ W)$

$1.2\ D + 1.6\ W + 0.5\ L + 0.5\ (L_r \text{ or } S)$

$0.9\ D + 1.6\ W$

Where:

- D... is the dead load or the weight of the structure itself
- L... is the live load which vary for different structures
- S... is the snow load
- W... is the wind load

The central problem of designing a member is to find a cross section for which the required strength doesn't exceed the available strength:

$P_u < ¢ P_n$ where P_u is the sum of the factored loads.

to prevent yielding

$0.90 F_y A_g > P_u$

to avoid fracture,

$0.75 F_u A_e > P_u$

Therefore, to design a tension member, we have to consider the loads applied to this member, the design forces acting on this member (M_u, P_u, and V_u), and the point where this member would fail.

In the design of a tension member, the design tensile force is given and the type of member and the size of the member have to be arrived at. The type of member is usually dictated by the location where the member is used. In the case of roof trusses, for example, angles or pipes are commonly used. Depending upon the span of the truss, the location of the member in the truss and the force in the member either single angle or double angles may be used in roof trusses. Single angle is common in the web members and the double angles are common in rafter and tie members of a roof truss. Plate tension members are used to suspend pipes and building floors. Rods are also used as suspenders and as sag rods of roof purlins. Steel wires are used as suspender cables in bridges and buildings. Pipes are used in roof trusses on aesthetic considerations, in spite of fabrication difficulty and the higher cost of such tubular trusses. Built-up members made of angles, channels and plates are used as heavy tension members, encountered in bridge trusses.

Trial and Error Design Process

The design process is iterative, involving choice of a trial section and analysis of

its capacity. This process is discussed in this section. Initially, the net effective area required is calculated from the design tension and the ultimate strength of the material as given below.

An = Ft / (fu / ³m1) *(4.12)*

Using the net area required, the gross area required is calculated, allowing for some assumed number and size of bolt holes in plates, or assumed efficiency index in the case of angles and threader rods. The gross area required is also checked against that required from the yield strength of the gross sections as given below.

Ag = Td / (fy / ³m0).

A suitable trial section is chosen from the steel section handbook to meet the

gross area required. The bolt holes are laid out appropriately in the member and the member is analyses to obtain the actual design strength of the trial section. The design strength of the trial section is evaluated using Eqs.4.1 to 4.5 in the case of plates and threaded bars and using Eqs.4.6 in the case of angle ties. If the actual design strength is smaller than or too large compared to the design force, a new trial section is chosen and the analysis is repeated until a satisfactory design is obtained.

Stiffness Requirement

The tension members, in addition to meeting the design strength requirement,

frequently have to be checked for adequate stiffness. This is done to ensure that the member does not sag too much during service due to self-weight or the eccentricity of end plate connections. The IS: 800 imposes the following limitations on the slenderness ratio of members subjected to tension:

(a) In the case of members that are normally under tension but may experience compression due to stress reversal caused by wind / earthquake loading $l / r \leq 250$.

(b) In the case of members that are designed for tension but may experience stress reversal for which it is not designed (as in X bracings). l / r d– 350

(c) In the case of members subjected to tension only. l / r d– 400

In the case of rods used as a tension member in X bracings, the slenderness ratio limitation need not be checked for if they are pre-tensioned by using a turnbuckle or other such arrangement.

It is now time to actually learn how to design something! We start with tension members because they are relatively simple. There are only a few limit states to worry about. Now that we are ready, please turn to Chapter D of the AISC specification (page 16.1-26) and let's get started.

Controlling Limit States

In SCM Chapter D we look at three limit states that relate to the member itself. We will then look to SCM Chapter J to look at some connection related limit states at apply to connecting elements and/or end conditions of general tension members.

The SCM Chapter D limit states that we will consider are:

- Slenderness
- Tensile yielding
- Tensile rupture

In SCM Chapter J we will look at the limit states of:

- Tensile yielding of connecting elements
- Tensile rupture of connecting elements
- Bolt bearing on the edge of the bolt hole
- Block shear rupture of strength at end connections of tension members

As discussed in an earlier chapter, limit states represent conditions that limit the usefulness of a member. Generally only one of the limit states will control the design. All must be checked to ensure that your member is adequate for its intended purpose. The following sections present the limit states as implemented by the AISC specification. The example problems show how the limits are applied.

Note that the connecting plate shown is a separate member and has its own set of limit states that used to define it's tensile strength. For this example, we will only focus on the capacity of the W section.

This failure mode looks at yielding on the gross cross sectional area, A_g, of the member under consideration. Consequently, the critical area is located away from the connection as shown. Strength of the section equals the *gross area*, A_g, times the *minimum yield stress*, F_y, of the member.

Tensile rupture occurs in the next section of the W section at the connection. In this case we have two potential failure paths that see the full force of the member. It is common to have multiple potential failure paths. Each valid path must be investigated. Tensile rupture is complicated by the need to get the forces out of the flanges, through the web, and into the bolts. This means that we need to account for the stress concentrated in and around the bolts. This will be discussed in further detail later. The capacity of each failure path equals the *effective net area*, A_e, times the *tensile stress*, F_u, of the member.

Block shear occurs when a "block" of the member is "torn" out as depicted. Block shear is characterised by a failure that includes both tension (i.e. normal to the force) and shear (i.e. parallel to the force) failure planes.

Like tensile rupture, there are frequently multiple valid failure paths that must be investigated. The capacity of each failure path is a sum of the capacities of each of the failure surfaces in the path. Each tension area capacity equals the tension area (either gross or net) times a tensile stress (yield or ultimate). Each shear area capacity equals the shear area (either gross or net) times a shear stress (yield or ultimate).

The three strength based limit states shown here are three of the four possible failures to be considered for the W section. Bolt bearing, the fourth strength limit state, is not reasonably shown in these figures and is treated later. The limit state that results in the lowest capacity for the member controls the *capacity of the member*.

Note that the *capacity of the connection* is the lowest of the capacity of the two connected members or the fasteners. This chapter deals with the capacity of the connected members, not the fasteners.

3

Design of Roof Trusses

Truss

In architecture and structural engineering, a truss is a structure comprising one or more triangular units constructed with straight members whose ends are connected at joints referred to as nodes. External forces and reactions to those forces are considered to act only at the nodes and result in forces in the members which are either tensile or compressive forces. Moments (torques) are explicitly excluded because, and only because, all the joints in a truss are treated as revolutes.

A planar truss is one where all the members and nodes lie within a two dimensional plane, while a space truss has members and nodes extending into three dimensions.

Characteristics of Trusses

A truss consists of straight members connected at joints. Trusses are composed of triangles because of the structural stability of that shape and design. A triangle is the simplest geometric figure that will not change shape when the lengths of the sides are fixed. In comparison, both the angles and the lengths of a four-sided figure must be fixed for it to retain its shape.

Planar Truss

The simplest form of a truss is one single triangle. This type of truss is seen in a framed roof consisting of rafters and a ceiling joist, and in other mechanical structures such as bicycles and aircraft. Because of the stability of this shape and the methods of analysis used to calculate the forces within it, a truss composed entirely of triangles is known as a simple truss. The traditional diamond-shape bicycle frame, which utilises two conjoined triangles, is an example of a simple truss. A planar truss lies in a single plane. Planar trusses are typically used in parallel to form roofs and bridges.

The depth of a truss, or the height between the upper and lower chords, is what makes it an efficient structural form. A solid girder or beam of equal strength would have substantial weight and material cost as compared to a truss. For a given span length, a deeper truss will require less material in the chords and greater material in the verticals and diagonals. An optimum depth of the truss will maximise the efficiency.

Space Frame Truss

A space frame truss is a three-dimensional framework of members pinned at their ends. A tetrahedron shape is the simplest space truss, consisting of six members which meet at four joints. Large planar structures may be composed from tetrahedrons with common edges and they are also employed in the base structures of large free-standing power line pylons

- Simple tetrahedron
- Diagram of a planar space frame such as used for a roof
- Four tetrahedrons form each of the two lower base structures of this power pylon

Truss Types

There are two basic types of truss:

- The *pitched truss*, or *common truss*, is characterised by its triangular shape. It is most often used for roof construction. Some common trusses are named according to their *web configuration*. The chord size and web configuration are determined by *span, load* and *spacing*.
- The *parallel chord truss*, or *flat truss*, gets its name from its parallel top and bottom chords. It is often used for floor construction.

A combination of the two is a truncated truss, used in hip roof construction. A metal plate-connected wood truss is a roof or floor truss whose wood members are connected with metal connector plates.

Pratt Truss

The Pratt truss was patented in 1844 by two Boston railway engineers, Caleb Pratt and his son Thomas Willis Pratt. The design uses vertical members for compression and horizontal members to respond to tension. What is remarkable about this style is that it remained popular even as wood gave way to iron, and even still as iron gave way to steel. The continued popularity of the Pratt truss is probably due to the fact that the configuration of the members means that longer diagonal members are only in tension for gravity load effects. This allows these members to be used more efficiently, as slenderness effects related to buckling under compression loads (which are compounded by the length of the member) will typically not control the design. Therefore, for given planar truss with a fixed depth, the Pratt configuration is usually the most efficient under static, vertical loading.

The Southern Pacific Railroad bridge in Tempe, Arizona is a 393 meter (1,291 foot) long truss bridge built in 1912. The structure is composed of nine Pratt truss spans of varying lengths. The bridge is still in use today.

The Wright Flyer used a Pratt truss in its wing construction, as the minimisation of compression member lengths allowed for lower aerodynamic drag.

Bowstring Truss

Named for their shape, bowstring trusses were first used for arched truss bridges, known as tied-arch bridges.

Thousands of bowstring trusses were used during World War II for holding up the curved roofs of aircraft hangars and other military buildings. Many variations exist in the arrangements of the members connecting the nodes of the upper arc with those of the lower, straight sequence of members, from nearly isosceles triangles to a variant of the Platt truss.

King Post Truss

One of the simplest truss styles to implement, the king post consists of two angled supports leaning into a common vertical support.

The queen post truss, sometimes *queenpost* or *queenspost*, is similar to a king post truss in that the outer supports are angled towards the center of the structure. The primary difference is the horizontal extension at the centre which relies on beam action to provide mechanical stability. This truss style is only suitable for relatively short spans.

Lenticular Truss

Lenticular trusses, patented in 1878 by William Douglas, have the top and bottom chords of the truss arched, forming a lens shape.

A lenticular pony truss bridge is a bridge design that involves a lenticular truss extending above and below the roadbed.

Town's Lattice Truss

American architect Ithiel Town designed Town's Lattice Truss as an alternative to heavy-timber bridges. His design, patented in 1820 and 1835, uses easy-to-handle planks arranged diagonally with short spaces in between them.

Vierendeel Truss

Figure: *A Vierendeel bridge; note the lack of diagonal elements in the primary structure and the way bending loads are carried between elements*

The Vierendeel truss is a truss where the members are not triangulated but form rectangular openings, and is a frame with fixed joints that are capable of transferring and resisting bending moments. Regular trusses comprise members that are commonly assumed to have pinned joints, with the implication that no moments exist at the jointed ends. This style of truss was named after the Belgian engineer Arthur Vierendeel, who developed the design in 1896. Its use for bridges is rare due to higher costs compared to a triangulated truss.

The utility of this type of truss in buildings is that a large amount of the exterior envelope remains unobstructed and can be used for fenestration and door openings. This is preferable to a braced-frame system, which would leave some areas obstructed by the diagonal braces.

Statics of Trusses

A truss that is assumed to comprise members that are connected by means of pin joints, and which is supported at both ends by means of hinged joints or rollers, is described as being statically determinate. Newton's Laws apply to the structure as a whole, as well as to each node or joint. In order for any node that may be subject to an external load or force to remain static in space, the following conditions must hold: the sums of all (horizontal and vertical) forces, as well as all moments acting about the node equal zero. Analysis of these conditions at each node yields the magnitudee may be compression or tension forces.

Trusses that are supported at more than two positions are said to be statically indeterminate, and the application of Newton's Laws alone is not sufficient to determine the member forces.

In order for a truss with pin-connected members to be stable, it must be entirely composed of triangles. In mathematical terms, we have the following necessary condition for stability:

$$m \geq 2j - r$$

where m is the total number of truss members, j is the total number of joints and r is the number of reactions (equal to 3 generally) in a 2-dimensional structure.

When m = 2j " 3, the truss is said to be *statically determinate*, because the (m+3) internal member forces and support reactions can then be completely determined by 2j equilibrium equations, once we know the external loads and the geometry of the truss. Given a certain number of joints, this is the minimum number of members, in the sense that if any member is taken out (or fails), then the truss as a whole fails. While the relation (a) is necessary, it is not sufficient for stability, which also depends on the truss geometry, support conditions and the load carrying capacity of the members.

Some structures are built with more than this minimum number of truss members. Those structures may survive even when some of the members fail. Their member forces depend on the relative stiffness of the members, in addition to the equilibrium condition described.

Analysis of Trusses

Because the forces in each of its two main girders are essentially planar, a truss is usually modelled as a two-dimensional plane frame. If there are significant out-of-plane forces, the structure must be modelled as a three-dimensional space.

The analysis of trusses often assumes that loads are applied to joints only and not at intermediate points along the members. The weight of the members is often insignificant compared to the applied loads and so is often omitted. If required, half of the weight of each member may be applied to its two end joints.

Provided the members are long and slender, the moments transmitted through the joints are negligible and they can be treated as "hinges" or 'pin-joints'.

Every member of the truss is then in pure compression or pure tension – shear, bending moment, and other more complex stresses are all practically zero. This makes trusses easier to analyse. This also makes trusses physically stronger than other ways of arranging material – because nearly every material can hold a much larger load in tension and compression than in shear, bending, torsion, or other kinds of force.

Structural analysis of trusses of any type can readily be carried out using a matrix method such as the direct stiffness method, the flexibility method or the finite element method.

Forces in Members

On the right is a simple, statically determinate flat truss with 9 joints and (2 x 9) " 3 = 15 members. External loads are concentrated in the outer joints. Since this is a symmetrical truss with symmetrical vertical loads, it is clear to see that the reactions at A and B are equal, vertical and half the total load.

The internal forces in the members of the truss can be calculated in a variety of ways including the graphical methods:

- *Cremona diagram*
- *Culmann diagram*
- *the analytical* Ritter *method (method of sections).*

Design of Members

A truss can be thought of as a beam where the web consists of a series of separate members instead of a continuous plate. In the truss, the lower horizontal member (the bottom chord) and the upper horizontal member (the top chord) carry tension and compression, fulfilling the same function as the flanges of an I-beam. Which chord carries tension and which carries compression depends on the overall direction of bending. In the truss pictured above right, the bottom chord is in tension, and the top chord in compression.

The diagonal and vertical members form the truss web, and carry the shear force. Individually, they are also in tension and compression, the exact arrangement of forces is depending on the type of truss and again on the direction of bending. In the truss shown above right, the vertical members are in tension, and the diagonals are in compression.

In addition to carrying the static forces, the members serve additional functions of stabilising each other, preventing buckling. In the picture to the right, the top chord is prevented from buckling by the presence of bracing and by the stiffness of the web members. The inclusion of the elements shown is largely an engineering decision based upon economics, being a balance between the costs of raw materials, off-site fabrication, component transportation, on-site erection, the availability of machinery and the cost of labor. In other cases the appearance of the structure may take on greater importance and so influence the design decisions beyond mere matters of economics. Modern materials such as prestressed concrete and fabrication methods, such as automated welding, have significantly influenced the design of modern bridges.

Once the force on each member is known, the next step is to determine the cross section of the individual truss members. For members under tension the cross-sectional area A can be found using $A = F \times \gamma / \sigma_y$, where F is the force in the member, γ is a safety factor (typically 1.5 but depending on building codes) and σ_y is the yield tensile strength of the steel used.

The members under compression also have to be designed to be safe against buckling.

The weight of a truss member depends directly on its cross section—that weight partially determines how strong the other members of the truss need to be.

Giving one member a larger cross section than on a previous iteration requires giving other members a larger cross section as well, to hold the greater weight of the first member—one needs to go through another iteration to find exactly how much greater the other members need to be.

Sometimes the designer goes through several iterations of the design process to converge on the "right" cross section for each member. On the other hand, reducing the size of one member from the previous iteration merely makes the other members have a larger (and more expensive) safety factor than is technically necessary, but doesn't *require* another iteration to find a buildable truss.

The effect of the weight of the individual truss members in a large truss, such as a bridge, is usually insignificant compared to the force of the external loads.

Design of Joints

After determining the minimum cross section of the members, the last step in the design of a truss would be detailing of the bolted joints, e.g., involving shear of the bolt connections used in the joints. Based of the needs of the project, truss internal connections (joints) can be designed as rigid, semi rigid, or hinged. Rigid connections can allow transfer of bending moments leading to development of secondary bending moments in the members.

Applications

Post Frame Structures

Component connections are critical to the structural integrity of a framing system. In buildings with large, clearspan wood trusses, the most critical connections are those between the truss and its supports. In addition to gravity-induced forces (a.k.a. bearing loads), these connections must resist shear forces acting perpendicular to the plane of the truss and uplift forces due to wind. Depending upon overall building design, the connections may also be required to transfer bending moment.

Wood posts enable the fabrication of strong, direct, yet inexpensive connections between large trusses and walls. Exact details for post-to-truss connections vary from designer to designer, and may be influenced by post type.

Solid-sawn timber and glulam posts are generally notched to form a truss bearing surface. The truss is rested on the notches and bolted into place. A special plate/bracket may be added to increase connection load transfer capabilities. With mechanically-laminated posts, the truss may rest on a shortened outer-ply or on a shortened inner-ply. The later scenario places the bolts in double shear and is a very effective connection.

Design

The nature of a truss allows for the analysis of the structure using a few assumptions and the application of Newton's laws of motion according to the branch of physics known as statics. For purposes of analysis, trusses are assumed to be pin jointed where the straight components meet. This assumption means that members of the truss (chords, verticals and diagonals) will act only in tension or compression. A more complex analysis is required where rigid joints impose significant bending loads upon the elements, as in a Vierendeel truss.

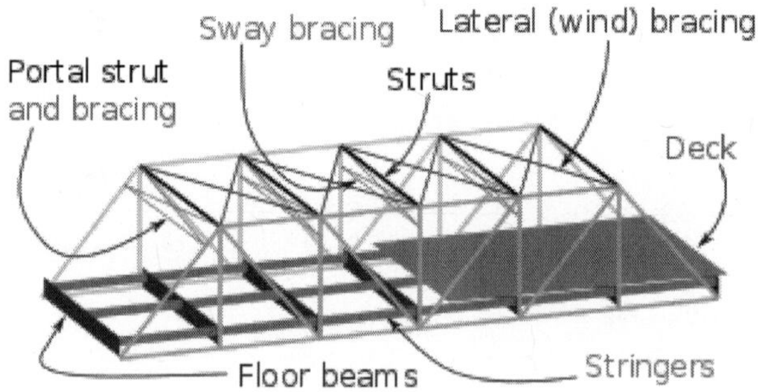

Figure: *The integral members of a truss bridge*

In the bridge illustrated in the infobox at the top, vertical members are in tension, lower horizontal members in tension, shear, and bending, outer diagonal and top members are in compression, while the inner diagonals are in tension. The central vertical member stabilises the upper compression member, preventing it from buckling. If the top member is sufficiently stiff then this vertical element may be eliminated. If the lower chord (a horizontal member of a truss) is sufficiently resistant to bending and shear, the outer vertical elements may be eliminated, but with additional strength added to other members in compensation. The ability to distribute the forces in various ways has led to a large variety of truss bridge types. Some types may be more advantageous when wood is employed for compression elements while other types may be easier to erect in particular site conditions, or when the balance between labor, machinery and material costs have certain favourable proportions.

The inclusion of the elements shown is largely an engineering decision based upon economics, being a balance between the costs of raw materials, off-site fabrication, component transportation, on-site erection, the availability of machinery and the cost of labor. In other cases the appearance of the structure may take on greater importance and so influence the design decisions beyond mere matters of economics. Modern materials such as prestressed concrete and fabrication methods, such as automated welding, and the changing price of steel relative to that of labor have significantly influenced the design of modern bridges.

History in the United States

Figure: *Warren-type through-truss of the former Seaboard Air Line Railway, located near the village of Willow, Florida; abandoned since the mid-1980s*

Because wood was so abundant, early truss bridges would typically use carefully fitted timbers for members taking compression and iron rods for tension members, usually constructed as a covered bridge to protect the structure. In 1820 a simple form of truss, Town's lattice truss, was patented, and had the advantage of not requiring high labor skills nor much metal. Few iron truss bridges were built in the United States before 1850.

Truss bridges became a common type of bridge built from the 1870s through the 1930s. Examples of these bridges still remain across the United States, but their numbers are dropping rapidly, as they are demolished and replaced with new structures. As metal slowly started to replace timber, wrought iron bridges in the U.S. started being built on a large scale in the 1870s. Bowstring truss bridges were a common truss design seen during this time, with their arched top chords.

Companies like the Massillon Bridge Company of Massillon, Ohio and the King Bridge Company of Cleveland, Ohio became well-known companies, as they marketed their designs to different cities and townships. The bowstring truss design (photo) fell out of favour due to a lack of durability, and gave way to the Pratt truss design, which was stronger. Again, the bridge companies marketed their designs, with the Wrought Iron Bridge Company in the lead. As the 1880s and 1890s progressed, steel began to replace wrought iron as the preferred material. Other truss designs were used during this time, including the camel-back. By the 1910s, many states developed standard plan truss bridges, including steel Warren pony truss bridges. As the 1920s and 1930s progressed, some states, such as Pennsylvania, continued to build steel truss bridges, including massive steel through truss bridges for long spans. Other states, such as Michigan, used standard plan concrete girder and beam bridges, and only a limited number of truss bridges were built.

Roadbed Types

The truss may carry its roadbed on top, in the middle, or at the bottom of the truss. Bridges with the roadbed at the top or the bottom are the most common as this allows both the top and bottom to be stiffened, forming a box truss. When the roadbed is atop the truss it is called a deck truss (an example of this was the I-35W Mississippi River bridge), when the truss members are both above and below the roadbed, it is called a through truss (an example of this application is the Pulaski Skyway), and where the sides extend above the roadbed but are not connected, a pony truss or half-through truss.

Sometimes both the upper and lower chords support roadbeds, forming a *double-decked truss.* This can be used to separate rail from road traffic or to separate the two directions of automobile traffic and so avoiding the likelihood of head-on collisions.

Truss Types Used in Bridges

Bridges are many times the best visible examples of truss use to the common person. There are many types of designs, many dating back hundreds of years. Below are some of the more common types and designs.

Allan Truss

The Allan Truss, designed by Percy Allan, is partly based on the Howe truss. The first Allan truss was completed on 13 August 1894 over Glennies Creek at Camberwell, New South Wales and the last Allan truss bridge was built over Mill Creek near Wisemans Ferry in 1929.

Completed in March 1895, the Tharwa Bridge located at Tharwa, Australian Capital Territory, was the second Allan truss to be built, the oldest surviving bridge in the Australian Capital Territory and the oldest, longest in continuous Allan truss bridge. Completed in November 1895, the Hampden Bridge in Wagga Wagga, New South Wales, Australia, the first of the Allan truss bridges with overhead bracing, was originally designed as a steel bridge but was constructed with timber to reduce cost.

In his design, Allan used Australian ironbark for its strength. A similar bridge also designed by Percy Allen is the Victoria Bridge on Prince Street Picton, New South Wales. Also constructed of ironbark, the bridge is still in use today for pedestrian and light traffic.

Bailey Bridge

Designed for military use, the prefabricated and standardised truss elements may be easily combined in various configurations to adapt to the needs at the site. In the image at right, note the use of doubled prefabrications to adapt to the span and load requirements. In other applications the trusses may be stacked vertically.

Baltimore Truss

The Baltimore truss is a subclass of the Pratt truss. A Baltimore truss has additional bracing in the lower section of the truss to prevent buckling in the compression members and to control deflection. It is mainly used for train bridges, boasting a simple and very strong design.

Bollman Truss

The Bollman Truss Railroad Bridge at Savage, Maryland is the only surviving example of a revolutionary design in the history of American bridge engineering. The type was named for its inventor, Wendel Bollman, a self-educated Baltimore engineer. It was the first successful all-metal bridge design (patented in 1852) to be adopted and consistently used on a railroad. The design employs wrought iron tension members and cast iron compression members. The use of multiple independent tension elements reduces the likelihood of catastrophic failure. The structure was also easy to assemble.

The Wells Creek Bollman Bridge is the only other bridge designed by Wendel Bollman still in existence, but it is a Warren truss configuration.

Bowstring Arch Truss (Tied Arch Bridge)

The bowstring arch through truss bridge was patented in 1841 by Squire Whipple. Thrust arches transform their vertical loads into a thrust along the arc of the arch. At the ends of the arch this thrust (at a downward angle away from the center of the bridge) may be resolved into two components, a vertical thrust equal to a proportion of the weight and load of the bridge section, and a horizontal thrust.

In a typical arch this horizontal thrust is taken into the ground, while in a bowstring arch the thrust is taken horizontally by a chord member to the opposite side of the arch. This allows the footings to take only vertical forces, useful for bridge sections resting upon high pylons.

Figure: *A tied arch bridge, in Pittsburgh, Pennsylvania. Note the stone pier in the background from the Wabash Bridge.*

Brown Truss

This type of truss is particularly suited for timber structures that use iron rods as tension members.

Burr Arch Truss

This combines an arch with a truss to form a structure both strong and rigid.

Cantilevered Truss

Most trusses have the lower chord under tension and the upper chord under compression. In a cantilever truss the situation is reversed, at least over a portion of the span. The typical cantilever truss bridge is a *balanced cantilever*, which enables the construction to proceed outward from a central vertical spar in each direction. Usually these are built in pairs until the outer sections may be anchored to footings. A central gap, if present, can then be filled by lifting a conventional truss into place or by building it in place using a *travelling support.*

Fink Truss

The Fink truss was designed by Albert Fink of Germany in the 1860s. This type of bridge was popular with the Baltimore and Ohio Railroad. The Appomattox High Bridge on the Norfolk and Western Railroad included 21 Fink deck truss spans from 1869 until their replacement in 1886.

Howe Truss

The relatively rare Howe truss, patented in 1840 by Massachusetts millwright William Howe, includes vertical members and diagonals that slope up towards the center, the opposite of the Pratt truss. In contrast to the Pratt Truss, the diagonal web members are in compression and the vertical web members are in tension. Examples include Jay Bridge in Jay, New York, and Sandy Creek Covered Bridge in Jefferson County, Missouri.

Figure: *Jay Bridge showing the truss design*

A large timber Howe truss in a commercial building

Howe truss illustrated - the diagonals are under compression under balanced loading

King Post Truss

K Truss

A truss in the form of a K due to the orientation of the vertical member and two oblique members in each panel.

Kingpost Truss

One of the simplest truss styles to implement, the king post consists of two angled supports leaning into a common vertical support.

Lattice Truss (Town's Lattice Truss)

This type of bridge uses a substantial number of lightweight elements, easing the task of construction. Truss elements are usually of wood, iron, or steel.

Lenticular Truss

A lenticular truss bridge includes a lens-shape truss, with trusses between an upper arch that curves up and then down to end points, and a lower arch that curves down and then up to meet at the same end points. Where the arches extend above and below the roadbed, it is a lenticular pony truss bridge. One type of lenticular truss consists of arcuate upper compression chords and lower eyebar chain tension links. The Royal Albert Bridge (United Kingdom) uses a single tubular upper chord. As the horizontal tension and compression forces are balanced these horizontal forces are not transferred to the supporting pylons (as is the case with most arch types). This in turn enables the truss to be fabricated on the ground and then to be raised by jacking as supporting masonry pylons are constructed. This truss has been used in the construction of a stadium, with the upper chords of parallel trusses supporting a roof that may be rolled back. The Smithfield Street Bridge in Pittsburgh, Pennsylvania is another example of this type.

An example of a lenticular pony truss bridge that uses regular spans of iron is the Turn-of-River Bridge designed and manufactured by the Berlin Iron Bridge Co..

Long Truss

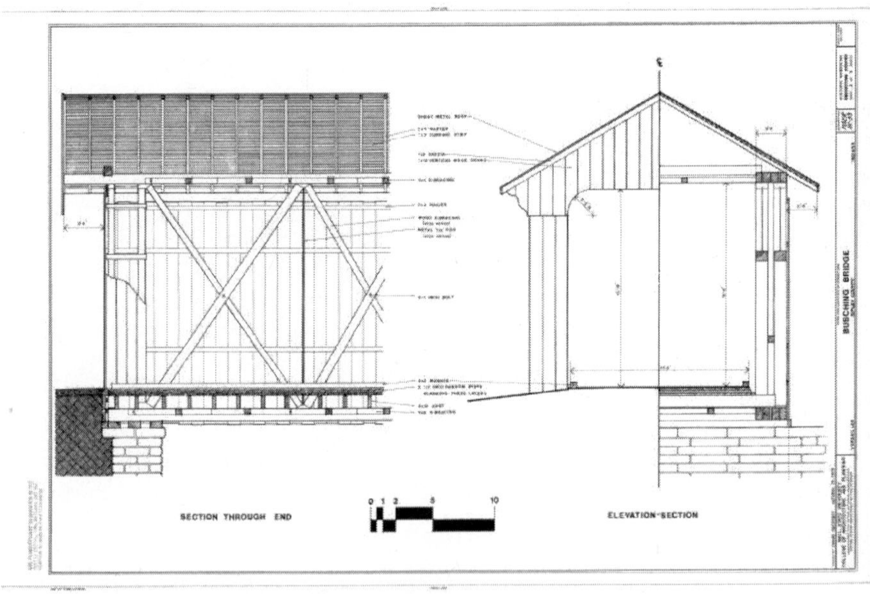

Figure: *HAER diagram of a Long Truss*

Designed by Stephen H. Long in 1830. The design resembles a Howe truss, but is entirely made of wood instead of a combination of wood and metal.

The longest surviving example is the Eldean Covered Bridge north of Troy, Ohio, spanning 224 feet. One of the earliest surviving examples is the Old Blenheim Bridge, which with a span of 210 feet and a total length of 232 feet long is the second-longest covered bridge in the United States.

Parker (Camelback) Truss

A Parker truss bridge is a Pratt truss design with a polygonal upper chord. A "camelback" is a subset of the Parker type, where the upper chord consists of exactly five segments. An example of a Parker truss is the Traffic Bridge in Saskatoon, Canada. An example of a camelback truss is the Woolsey Bridge near Woolsey, Arkansas.

Pegram Truss

The Pegram truss is a hybrid between the Warren and Parker trusses where the upper chords are all of equal length and the lower chords are longer than the corresponding upper chord. Because of the difference in upper and lower chord length, each panel was not square. The members which would be vertical in a Parker truss vary from near vertical in the center of the span to diagonal near each end (like a Warren truss). George H. Pegram, while the chief engineer of Edge Moor Iron Company in Wilmington, Delaware, patented this truss design in 1885. The Pegram truss consists of a Parker type design with the vertical posts leaning towards the center at an angle between 60 and 75°. The variable post angle and constant chord length allowed steel in existing bridges to be recycled into a new span using the Pegram truss design. This design also facilitated reassembly and permitted a bridge to be adjusted to fit different span lengths. There are ten remaining Pegram span bridges in the United States with seven in Idaho.

Pennsylvania (Petit) Truss

The Pennsylvania (Petit) truss is a variation on the Pratt truss. Examples of this truss type include the Schell Bridge in Northfield, Massachusetts and the Inclined Plane Bridge in Johnstown, Pennsylvania.

Post Truss

A Post truss is a hybrid between a Warren truss and a double-intersection Pratt truss. Invented in 1863 by Simeon S. Post, it is occasionally referred to as a *Post patent truss* although he never received a patent for it. The Ponakin Bridge and the Bell Ford Bridge are two examples of this truss.

Pratt Truss

A Pratt truss includes vertical members and diagonals that slope down towards the center, the opposite of the Howe truss. It can be subdivided, creating Y- and K-shaped patterns. The Pratt Truss was invented in 1844 by Thomas and Caleb Pratt. This truss is practical for use with spans up to 250 feet and was a common configuration for railroad bridges as truss bridges moved from wood to metal. They are statically determinate bridges, which lend themselves well to long spans.

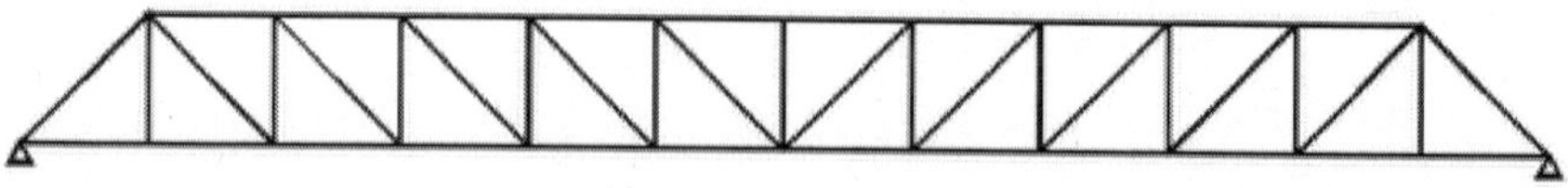

Figure: *Pratt truss illustrated - the interior diagonals are under tension under balanced loading and vertical elements under compression. If pure tension elements are used in the diagonals (such as eyebars) then crossing elements may be needed near the center to accept concentrated live loads as they traverse the span.*

Queenpost Truss

The queenpost truss, sometimes *queen post* or *queenspost*, is similar to a king post truss in that the outer supports are angled towards the center of the structure. The primary difference is the horizontal extension at the center which relies on beam action to provide mechanical stability. This truss style is only suitable for relatively short spans.

Truss Arch

Eastbound Canadian Pacific Railway freight train over the truss-arch Stoney Creek Bridge on the Mountain Subdivision near Rogers Pass, British Columbia Canada. A truss arch may contain all horizontal forces within the arch itself, or alternatively may be either a thrust arch consisting of a truss, or of two arcuate sections pinned at the apex. The latter form is common when the bridge is constructed as cantilever segments from each side as in the Navajo Bridge.

Waddell Truss

Patented 1894 (U.S. Patent 529,220) its simplicity eases erection at the site. It was intended to be used as a railroad bridge.

Warren Truss

The Warren truss was patented in 1848 by its designers James Warren and Willoughby Theobald Monzani, and consists of longitudinal members joined only by angled cross-members, forming alternately inverted equilateral triangle-shaped spaces along its length, ensuring that no individual strut, beam, or tie is subject to bending or torsional straining forces, but only to tension or compression. Loads on the diagonals alternate between compression and tension (approaching the center), with no vertical elements, while elements near the center must support both tension and compression in response to live loads. This configuration combines strength with economy of materials and can therefore be relatively light. It is an improvement over the Neville truss which uses a spacing configuration of isosceles triangles.

Figure: *A preserved original Ansaldo SVA aircraft, showing the Warren truss-pattern interplane wing strut layout*

Warren truss construction has also been used in airframe construction for aircraft since the 1920s, mostly for smaller aircraft fuselages, using chrome molybdenum alloy steel tubing, with popular aircraft such as the Piper J-3 *Cub*. One of the earliest uses for the Warren truss design in aircraft design was for the interplane wing strut layout, as seen in a nose-on view, on the Italian World War I Ansaldo SVA series of fast reconnaissance biplanes, which were among the fastest aircraft of the First World War era. Warren truss construction is still used today for some homebuilt aircraft fuselage designs, that essentially use the same 1920s-era design philosophies in the 21st century.

Whipple Pratt Truss

A whipple truss is usually considered a subclass of the Pratt truss because the diagonal members are designed to work in tension. The main characteristic of a whipple truss is that the tension members are elongated, usually thin, at a shallow angle and cross two or more bays (rectangular sections defined by the vertical members).

Vierendeel Truss

The Vierendeel truss, unlike common pin-jointed trusses, imposes significant bending forces upon its members — but this in turn allows the elimination of many diagonal elements. While rare as a bridge type this truss is commonly employed in modern building construction as it allows the resolution of gross shear forces against the frame elements while retaining rectangular openings between columns. This is advantageous both in allowing flexibility in the use of the building space and freedom in selection of the building's outer curtain wall, which affects both interior and exterior styling aspects

Tied-arch bridge: Fremont Bridge in Portland, Oregon is the second-longest tied arch bridge in the world. In this case the half-arches on either end of the bridge terminate atop slender raised piers. The horizontal forces of the main arch and the outer half-arches balance at the ground level foundations.

A tied-arch bridge is an arch bridge in which the outward-directed horizontal forces of the arch, or top chord, are borne as tension by the bottom chord (either tie-rods or the deck itself), rather than by the ground or the bridge foundations.

Thrusts downward on such a bridge's deck are translated, as tension, by vertical ties of the deck to the curved top chord, tending to flatten it and thereby to push its tips outward into the abutments, like other arch bridges. However in a tied-arch or bowstring bridge, these movements are restrained not by the abutments but by the bottom chord, which ties these tips together, taking the thrusts as tension, rather like the string of a bow that is being flattened. Therefore the design is often called a bowstring-arch or bowstring-girder bridge. The elimination of horizontal forces at the abutments allows tied-arch bridges to be constructed with less robust foundations; thus they can be situated atop elevated piers or in areas of unstable soil. In addition, since they do not depend on horizontal compression forces for their integrity, tied-arch bridges can be prefabricated offsite, and subsequently floated, hauled or lifted into place. A notable case of these procedures was the installation of the Fremont Bridge in Portland, Oregon.

Issues

In a 1978 advisory issued by the Federal Highway Administration (FHWA), the FHWA noted that tied-arch bridges are susceptible to problems caused by poor welds at the connection between the arch rib and the tie girders, and at the connection between the arch and vertical ties. In addition, problems with electroslag welds, while not isolated to tied-arch bridges, resulted in costly, time-consuming and inconveniencing repairs. The structure as a whole was described as nonredundant; failure of either of the two tie girders would result in failure of the entire structure.

King Post

A so-called "Norman truss" in the 18th-century Bolduc House in Ste. Geneviève, Missouri

A king post (or king-post) is a central vertical supporting post used in architectural, bridge, or aircraft design applications.

Architecture

A king post (or crown post) extends vertically from a crossbeam to the apex of a triangular truss. The king post connects the apex of the truss with its base, holding up the tie beam at the base of the truss. King posts were used in roof construction in Medieval architecture in buildings such as parish churches and tithe barns, and also appear in Gothic Revival architecture and Queen Anne architecture. A similar structure may be used to construct a simple bridge. A development of the king post structure terminates the central post before it reaches the apex of the roof truss. Instead, the king post is held in place by a collar beam, a secondary beam parallel to the tie beam. In this structure, the central post is termed the crown post.

An alternative truss construction uses two queen posts (or queen-posts). These vertical posts, positioned along the base of the truss, are supported by the sloping sides of the truss, rather than reaching its apex. A development adds a collar beam above the queen posts, which are then termed queen struts. A section of the tie beam between the queen posts may be removed to create a hammerbeam roof.

Norman Truss

Architectural historians in Louisiana use the term "Norman roof" to refer to a steeply-pitched roof; it is supported by what they call a "Norman truss". This is a through-purlin truss consisting of a tie beam and paired truss blades, with a central king post to support the roof ridge. The name derives from a belief that this system of construction was introduced to North America by settlers from Normandy in northern France, but it is really a misnomer as the system was more widely used than that.

King Post Truss

The king post truss is used for simple roof trusses and short-span bridges. It is the simplest form of truss in that it is constructed of the fewest number of truss members (individual lengths of wood or metal). The truss consists of two diagonal members that meet at the apex of the truss, one horizontal beam that serves to tie the bottom end of the diagonals together, and the king post which connects the apex to the horizontal beam below. For a roof truss, the diagonal members are called rafters, and the horizontal member may serve as a ceiling joist. A bridge would require two king post trusses with the driving surface between them. A roof usually uses many side-by-side trusses depending on the size of the structure.

Marine Engineering

On a cargo ship a king post is an upright with cargo-handling devices attached to it. Since king posts are designed for handling cargo, they are located at the forward or after end of a hatch.

Brown Truss

The Brown truss is a box truss that is a through truss (as contrasted with a deck truss) and consists of diagonal cross compression members connected to horizontal top and bottom stringers. There may be vertical or almost vertical tension members (the diagram shows these members, while the patent application diagram does not) but there are no vertical members in compression. In practice, when used in a covered bridge, the most common application, the truss is protected with outside sheathing.

The floor and roof are also trusses, but are horizontal and serve to give the truss rigidity. The bottoms of the diagonals tend to protrude below the sheathing. The Brown truss is noted for economy of materials as it can be built with very little metal.

Patent

Brown's patent claims did not actually address the economy afforded by lack of vertical members ("braces"). Instead he focused on the improved strength over previous trusses that had members ("braces" in his terminology) come to the horizontal chord near to each other but not exactly together (at "gains" in his terminology), by having several members come together in the same place. From the patent text:

I do not claim broadly furnishing the main or counter braces with gains and passing them between the timbers of the chords;

What I do claim as my invention, and desire to secure by letters Patent, is— Providing each of the main and counter braces with two gains at top and bottom, and each of the timbers of the chord with a gain at the point where the braces are applied corresponding with the gains in the braces, and the braces thus formed up between the timber, with the gains of the braces in such relation to the gains of the timbers that when the timbers of the chords are brought together they are combined and become, as it were, only one piece, no part of which can be operated upon or affected independently of the other by the downward and upward thrusts common to truss bridges, even if the bolt which passes laterally through and intersects each set of braces and the timbers of the chord were removed.

4

Fixed Connections

Fixed connections are very common. Steel structures of many sizes are composed of elements which are welded together. A cast-in-place concrete structure is automatically monolithic and it becomes a series of rigid connections with the proper placement of the reinforcing steel. Fixed connections demand greater attention during construction and are often the source of building failures.

Let this small chair illustrate the way in which two types of "fixed" connections can be generated. One is welded and the other is comprised to two screws. Both are considered to be fixed connections due to the fact that both of them can resist vertical and lateral loads as well as develop a resistance to moment. Thus, it found that not all fixed connections must be welded or monolithic in nature.

Simple Supports

Simple supports are idealized by some to be frictionless surface supports. This is correct in as much as the resulting reaction is always a single force that is perpendicular to, and away from, the surface. However, are also similar to roller supports in this. They are dissimilar in that a simple support cannot resist lateral loads of any magnitude. The built reality often depends upon gravity and friction to develop a minimal amount of frictional resistance to moderate lateral loading. For example, if a plank is laid across gap to provide a bridge, it is assumed that the plank will remain in its place.

It will do so until a foot kicks it or moves it. At that moment the plank will move because the simple connection cannot develop any resistance to the lateral loal. A simple support can be found as a type of support for long bridges or roof span. Simple supports are often found in zones of frequent seismic activity.

Implications

The following movies illustrate the implications of the type of support condition on the deflection behaviour and on the location of maximum bending stresses of a beam supported at its ends.

Drilled Hole and Bolt

This approach is probably the least desireable one for a strong reliable connection. It is commonly attempted by drilling a hole that is the same nominal size as the bolt. This is often done with a hand drill. This way of doing it will rarely, if ever, result in a tightly fitting connection that will remain in its original condition.

Here is why:

1) Drill bits rarely cut accurately sized cylindrical holes. The best that can be done is to use a new, or professionally resharpened bit in a heavy duty drill press or mill. The hole will be slightly larger than the nominal size. It is better for the bit to be ground to be center cutting in the area of the web, to eliminate the high cutting pressure required to push the web thru the material. If the bit is not center cutting, it will tend to wander around while trying to mash the material in its way out to the cutting edges on the face of the bit. This causes us to push harder, trying to force it through the material, causing it to wander more. There is a way to do a good drilling job with a non-centercutting drill, to be covered later.

   If the bit is worn, it probably does not have the same sharpness on each cutting edge on the tip. This will cause the bit to wander and cut a non-round hole as one edge grabs more material than the other during rotation. This causes it to snake its way through the material.

   Depending on drill condition, cutting fluid, and drilling speed and pressure, the chips coming out of the hole can cause some wear and scoring of the hole surface, and cause the starting end of the hole to get a little larger.

   A drill bit is not a very rigid cutter, it can flex quite a bit and enlarge and taper the hole, if not sharp and fed properly into the material.

2) Using an hand drill motor to drill the hole increase the errors in the finished hole. We can never hold it in perfect alignment

during the drilling, this will cause the hole size to increase. And, we can never be as rigid as the drill press or mill, so any tendency for the bit to bend and wander is greater, increasing hole size.

3) Bolts are usually not close tolerance pins. The ones we can get at the hardware store, or even a good industrial supply, do not have perfectly cylindrical shanks, and their size tolerance is relatively large for this application. Most bolts are meant to fit through holes with clearance and clamp things together, instead of being pins for transferring loads by bearing on the holes they are in. Close tolerance bolts can be obtained from aircraft suppliers from the "AN" line of fasteners. Another choice for a close tolerance bolt, would be precision ground shoulder bolts. Their grip length is important to get right.

All of this is not good for a connection that relies entirely on the pinning to hold it in place. What we usually get is a loosely fitting connection that has little of the available pinned surface area in intimate contact.

The slop in the connection allows movement, which results in higher dynamic loads on the mating faces of the cylindrical interface. Even when the surfaces come into contact, there is not full surface contact due to the irregularities, so the pressure on the contacting surfaces are much higher than expected.

This often results in material yielding, the deformation increases the looseness of the connection, which increase the dynamic loads developed, which results in more yielding....etc. A loose fitting pinned connection will usually only get worse.

The only way to make this type of connection remain stable is to (I.E. stay as loose as it is, not get worse) is to use large enough pins to reduce the bearing loads on the holes to a point where the material will not yield when the pin is banging back and forth on the hole edges. Using more pins will help.

Now, if we are only expecting the pinning to prevent slippage of a mast clamp during the relatively infrequent times when the clamp is overloaded, it will probably work for an acceptable period, possibly forever, if the clamp is not slipping very often. The clamp will reduce the overall load on the pinned connection and eliminate most of the dynamic load peaks applied to it.

Drilled and Reamed Hole with a Precision Straight Pin

To straight pin something, we need a precision ground shoulder bolt or dowel pin, a special reamer to get the hole sized for a press fit, and the mating parts need to be fixed when drilling and reaming the holes. The hole is drilled slightly undersize, then accurately sized with a reamer to obtain a close tolerance cylindrical hole. The reamer size is selected to establish the fit to the pin. To obtain a sound connection the hole should be just a bit smaller than the pin creating a "press fit". This ensures maximum contact between the pin and hole. A dowel pin can be used as the press fit will hold it in place.

Again, this is usually difficult to do properly with a hand drill, but will be much better than the drill & bolt approach.

A tried and true way to get a good pinned connection is to use "Morse" taper pins. They are available from any good industrial supply or your gear & bearing guy. The "Morse" taper is one big long taper (at 11 degrees, if I remember it right, the books not here) and is divided into different sections, designated by a number. You can find info in the Machinery Handbook, probably one of the most useful general references a gearhead can have, or see your supply guy's book.

The Morse" taper is a self-locking taper, which means that once the connection is seated, by tapping the pin in place, it will not easily come loose. You can thread the small end of the pin with a die, to install a self-locking nut, for extra insurance.

This connection requires a special tapered reamer to finish the holes in the mating parts, and the mating parts need to be fixed during the drilling and reaming process.

These things have been used for ages and work very well. Open up a piece of industrial machinery, and you will likely find some of them.

I helped a friend pin his mast inside a tube welded to a prop pitch with two taper pins at right angles to each other. It turned a 5 element 42' boom 20 and a 2 element

KLM 40 for years on a very windy hilltop location. Absolutely no problems.

When I was a youngster, an ex-pro mechanic friend and I built a 150 mph racing kart. All of the connections on the rear axle were made with small taper pins. We threaded the small ends and used lock nuts to keep them from vibrating loose. No, problems at all, worked great, and real easy to R&R during a panic.

Since, we need to buy a special pin and a special reamer to make a good pinned connection, I would opt for the tapered version because:

1) Making the tapered hole is easier to do properly. Drilling and reaming a constant diameter hole requires a constant stable alignment of the cutter. Wiggling the hand drill is not good, but rarely avoidable. The tapered reamer is pretty much a self aligning cutter. You may get the hole off axis a bit, but what the reamer cuts will be true to itself, and a true conical surface. That's what the pin needs to work properly.
2) The tapered pin is easier to R&R, and more durable during service. Just tap the pin to unlock the taper, and then it comes right out. The straight press fit pin has to be driven out for it's entire length. R&R'ing the straight pin causes surface wear, which means fewer R&R's before the connection needs to be redone.
3) The tapered pin connection can usually be renewed several times with the original tool investment. Just chase the hole with the reamer, and get a new pin (if required). When the straight pinned connection is renewed, it's a new reamer and pin for the next larger size.

Why Use Steel?

Steel construction has so many advantages: the strength to weight ratio is excellent, metals join easily, efficient shapes are available, etc. With those advantages, though, come some challenges that are best solved by a good understanding of how the metals actually perform in a structure.

For most larger buildings, metals are a key element of the structural system. Steel beams and columns, steel joists, steel studs, aluminium framing are a few examples of metal construction. A wise Superintendent understands not only how to erect the structure, but also the basics of how the structural system works and the ways in which it fails.

The Kansas City Hyatt Hotel structure spectacularly failed a few years ago, with a suspended walkway crashing to the ground. Many people were hurt and killed. The cause was traced to what seemed to be a minor change made during construction. This minor change, however, was structurally significant, increasing the shear stress well beyond design limits. The lesson learned from the Kansas City Hyatt Regency hotel collapse is that structural steel design, shop drawing review, fabrication and erection must be carefully done, with responsibilities clearly understood.

Do You Want a Brief History of Steel?

To understand metals, start with iron. As a basic chemical element Fe, iron is the most abundant metal on Earth. For many centuries, iron furnaces have heated limestone and iron ore that was excavated from the ground. The intense heat melts the both the rock and iron ore, along with several chemical reactions and the lighter liquid rock rises to the top and the heavier liquid iron sinks, creating pig iron. This pig iron is an intermediate step on the way to a final product.

Historically, wrought iron was a building product made from this pig iron. The wrought iron was mostly pure iron (with some slag and small amounts of carbon added). Wrought iron was actually "wrought" (i.e. worked or hammered) into bars and has been used as a construction material for thousands of years. Wrought iron is tough and ductile, easy to weld. Lacking the carbon content for tempering, wrought iron is not hard enough to hold a good edge for a tool or weapon.

Other final products from the pig iron are alloys. An alloy is a combination of two or more elements, in which at least one is a metal. Most metals used for construction purposes are alloys. For example, steel is an alloy with iron and carbon being the primary elements. Generally, iron-carbon alloys with up to 2.1% carbon by weight are considered steel and iron-carbon alloys with greater amounts of carbon are cast iron.

Cast iron is made by re-melting pig iron in a blast furnace, removing undesirable elements like phosphorus and sulfur, adjusting carbon levels and adding other elements. The resulting alloy, commonly called grey cast iron, has a high corrosion resistance and strong compressive strength, but tends to be brittle and difficult to weld. Historically, cannons and cannon balls were made from grey cast iron, as well as some early bridges. Steel is an alloy that finds tremendous number of uses in today's construction world. Hot rolled steel shapes, most commonly found as steel beams and columns on construction projects, are created in steel mills by rolling the heat steel between large rollers, deforming the steel into the typical shapes: W, S, C, angles, tube sections, pipes, etc. Most hot formed steel is either 36,000 psi or 50,000 psi yield strength.

Why are Connections So Important in Structural Steel?

Most structural steel failures happen at connections....where a beam connects to a column, where a joist connects to a beam, where a hanging rod connects to a beam (the Kansas City Hyatt discussed above). The Structural Engineer must design the design the steel members and give guidelines for the connections. Many people in the Construction Industry don't understand, though, that the Structural Engineer rarely designs the connections.

Why is that? Historically, the Steel Fabricators developed many different ways to make connections. What one Fabricator did in his shop economically might have been quite an expensive way to do it in a competitor's Fabrication Shop. So the practice developed that the Structural Engineer would size the members, but the Steel Fabricators would design the connections, which the Structural Engineer should then review and approve. If you think that seems like a complicated system prone to error, you'd be correct.

But that is the system we generally have in American construction. So the Construction Supervisor should know something about steel connections and have an idea if they are being installed correctly. A bit of background in Basic Structural Design is helpful, but the main thing to understand is the concept of pin connections versus fixed connections. A beam bolted to a column with clip angles along the beam web likely creates a pin connection. This means that the beam shouldn't be able to move up or down, nor in or out, but it can rotate a bit. A steel column bolted to a concrete pier with four anchor bolts also typically creates a pin connection. Again the steel column won't go up, down or sideways, but it may be able to rotate a bit.

The fixed connection must stop that ability to rotate. So for a beam to have a fixed connection to a column, along with clip angles, there may be a plate on the top and bottom flanges of the beam that gets welded to the column. With all that welding, the beam can no longer rotate. If a steel column is buried four feet deep in a concrete pier, it also would not be rotating at the point that it exits from the concrete. So those are a couple of ways to create fixed (or moment resisting) connections.

The Construction Supervisor should be aware if any fixed (or moment resisting) connections are required and understand how they are to be made. Just asking the questions increases the likelihood of a successful project.

What is Composite Design?

Composite design marries some of steel and concrete's best attributes together for an efficient structural system. Let's start by thinking about a structural system that isn't composite design. Structural steel beams placed at 4' on center with a steel deck spanning perpendicular which will have 4– of concrete placed on top of the steel deck is not a composite system. That means the steel beams will carry their own weight, the weight of the steel deck and concrete above and whatever live load gets applied. The steel deck and the concrete must carry their own weight and the live load spanning from steel beam to steel beam. Another way to state the proposition: the steel beam acts on its own structurally and the steel deck and concrete act on their own structurally.

A composite system ties together that steel beam and concrete floor and forces them to act as a single structural unit. Some connector on top of the steel beam makes the steel and concrete act as one unit. The steel beam can't slide independently of the concrete slab, the two are bonded together. Since the concrete is strong in compression, the composite system can be quite efficient structurally. The figure below illustrates the concept.

5

Seismic Loading

Seismic loading is one of the basic concepts of earthquake engineering which means application of an earthquake-generated agitation to a structure. It happens at contact surfaces of a structure either with the ground, or with adjacent structures, or with gravity waves from tsunami.

Seismic loading depends, primarily, on:

- Anticipated earthquake's parameters at the site - known as seismic hazard
- Geotechnical parameters of the site
- Structure's parameters
- Characteristics of the anticipated gravity waves from tsunami (if applicable).

Sometimes, seismic load exceeds ability of a structure to resist it without being broken, partially or completely. Due to their mutual interaction, seismic loading and seismic performance of a structure are intimately related.

Seismic Analysis

Seismic Analysis is a subset of structural analysis and is the calculation of the response of a building (or nonbuilding) structure to earthquakes. It is part of the process of structural design, earthquake engineering or structural assessment and retrofit in regions where earthquakes are prevalent.

As seen, a building has the potential to 'wave' back and forth during an earthquake (or even a severe wind storm). This is called the 'fundamental mode', and is the lowest frequency of building response. Most buildings, however, have higher modes of response, which are uniquely activated during earthquakes. The figure just shows the second mode, but there are higher 'shimmy' (abnormal vibration) modes. Nevertheless, the first and second modes tend to cause the most damage in most cases.

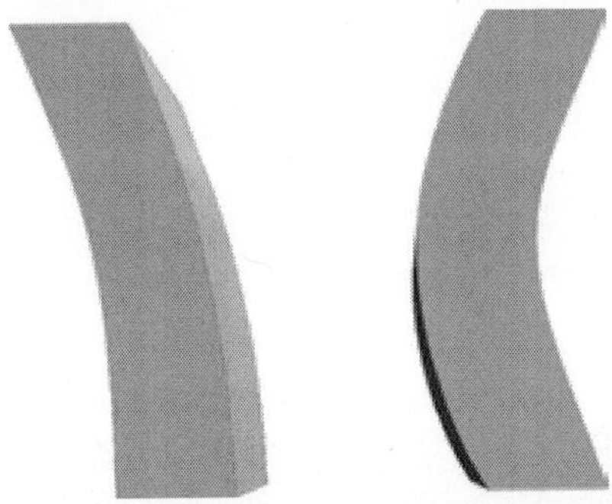

Figure: *First and second modes of building seismic response*

The earliest provisions for seismic resistance were the requirement to design for a lateral force equal to a proportion of the building weight (applied at each floor level). This approach was adopted in the appendix of the 1927 Uniform Building Code (UBC), which was used on the west coast of the USA.

It later became clear that the dynamic properties of the structure affected the loads generated during an earthquake. In the Los Angeles County Building Code of 1943 a provision to vary the load based on the number of floor levels was adopted (based on research carried out at Caltech in collaboration with Stanford University and the U.S. Coast and Geodetic Survey, which started in 1937). The concept of "response spectra" was developed in the 1930s, but it wasn't until 1952 that a joint committee of the San Francisco Section of the ASCE and the Structural Engineers Association of Northern California (SEAONC) proposed using the building period (the inverse of the frequency) to determine lateral forces.

The University of California, Berkeley was an early base for computer-based seismic analysis of structures, led by Professor Ray Clough (who coined the term finite element). Students included Ed Wilson, who went on to write the program SAP in 1970, an early "Finite Element Analysis" program.

Earthquake engineering has developed a lot since the early days, and some of the more complex designs now use special earthquake protective elements either just in the foundation (base isolation) or distributed throughout the structure. Analysing these types of structures requires specialised explicit finite element computer code, which divides time into very small slices and models the actual physics, much like common video games often have "physics engines". Very large and complex buildings can be modelled in this way (such as the Osaka Internation Convention Center).

Structural analysis methods can be divided into the following five categories.

Equivalent Static Analysis

This approach defines a series of forces acting on a building to represent the effect of earthquake ground motion, typically defined by a seismic design response spectrum. It assumes that the building responds in its fundamental mode. For this to be true, the building must be low-rise and must not twist significantly when the ground moves. The response is read from a design response spectrum, given the natural frequency of the building (either calculated or defined by the building code). The applicability of this method is extended in many building codes by applying factors to account for higher buildings with some higher modes, and for low levels of twisting. To account for effects due to "yielding" of the structure, many codes apply modification factors that reduce the design forces (e.g. force reduction factors).

Response Spectrum Analysis

This approach permits the multiple modes of response of a building to be taken into account (in the frequency domain). This is required in many building codes for all except for very simple or very complex structures. The response of a structure can be defined as a combination of many special shapes (modes) that in a vibrating string correspond to the "harmonics". Computer analysis can be used to determine these modes for a structure. For each mode, a response is read from the design spectrum, based on the modal frequency and the modal mass, and they are then combined to provide an estimate of the total response of the structure in this we have to calculate the magnitude of forces in all directions i.e X, Y & Z and then see the effects on the building.. Combination methods include the following:

- absolute - peak values are added together
- square root of the sum of the squares (SRSS)
- complete quadratic combination (CQC) - a method that is an improvement on SRSS for closely spaced modes

The result of a response spectrum analysis using the response spectrum from a ground motion is typically different from that which would be calculated directly from a linear dynamic analysis using that ground motion directly, since phase information is lost in the process of generating the response spectrum.

In cases where structures are either too irregular, too tall or of significance to a community in disaster response, the response spectrum approach is no longer appropriate, and more complex analysis is often required, such as non-linear static or dynamic analysis.

A sample response spectrum analysis.

Linear Dynamic Analysis

Static procedures are appropriate when higher mode effects are not significant. This is generally true for short, regular buildings. Therefore, for tall buildings, buildings with torsional irregularities, or non-orthogonal systems, a dynamic procedure is required. In the linear dynamic procedure, the building is modelled as a multi-degree-of-freedom (MDOF) system with a linear elastic stiffness matrix and an equivalent viscous damping matrix.

The seismic input is modelled using either modal spectral analysis or time history analysis but in both cases, the corresponding internal forces and displacements are determined using linear elastic analysis. The advantage of these linear dynamic procedures with respect to linear static procedures is that higher modes can be considered. However, they are based on linear elastic response and hence the applicability decreases with increasing nonlinear behaviour, which is approximated by global force reduction factors.

In linear dynamic analysis, the response of the structure to ground motion is calculated in the time domain, and all phase information is therefore maintained. Only linear properties are assumed. The analytical method can use modal decomposition as a means of reducing the degrees of freedom in the analysis.

Non-linear Static Analysis

In general, linear procedures are applicable when the structure is expected to remain nearly elastic for the level of ground motion or when the design results in nearly uniform distribution of nonlinear response throughout the structure. As the performance objective of the structure implies greater inelastic demands, the uncertainty with linear procedures increases to a point that requires a high level of conservatism in demand assumptions and acceptability criteria to avoid unintended performance. Therefore, procedures incorporating inelastic analysis can reduce the uncertainty and conservatism.

This approach is also known as "pushover" analysis. A pattern of forces is applied to a structural model that includes non-linear properties (such as steel yield), and the total force is plotted against a reference displacement to define a capacity curve. This can then be combined with a demand curve (typically in the form of an acceleration-displacement response spectrum (ADRS)). This essentially reduces the problem to a single degree of freedom (SDOF) system.

Nonlinear static procedures use equivalent SDOF structural models and represent seismic ground motion with response spectra. Story drifts and component actions are related subsequently to the global demand parameter by the pushover or capacity curves that are the basis of the non-linear static procedures.

Non-linear Dynamic Analysis

Nonlinear dynamic analysis utilises the combination of ground motion records with a detailed structural model, therefore is capable of producing results with relatively low uncertainty. In nonlinear dynamic analyses, the detailed structural model subjected to a ground-motion record produces estimates of component deformations for each degree of freedom in the model and the modal responses are combined using schemes such as the square-root-sum-of-squares.

In non-linear dynamic analysis, the non-linear properties of the structure are considered as part of a time domain analysis. This approach is the most rigorous, and is required by some building codes for buildings of unusual configuration or of special importance. However, the calculated response can be very sensitive to the characteristics of the individual ground motion used as seismic input; therefore, several analyses are required using different ground motion records to achieve a reliable estimation of the probabilistic distribution of structural response. Since the properties of the seismic response depend on the intensity, or severity, of the seismic shaking, a comprehensive assessment calls for numerous nonlinear dynamic analyses at various levels of intensity to represent different possible earthquake scenarios. This has led to the emergence of methods like the Incremental Dynamic Analysis.

Earthquake Resistant Structures

Earthquake engineering is treating its subject structures like defensive fortifications in military engineering but for the warfare on earthquakes. Both earthquake and military general design principles are similar: be ready to slow down or mitigate the advance of a possible attacker.

According to building codes, earthquake-resistant structures are meant to withstand the largest earthquake of a certain probability that is likely to occur at their location. This means the loss of life should be minimised by preventing collapse of the buildings for rare earthquakes while the loss of functionality should be limited for more frequent ones.

Ancient architects believed that devastating earthquakes were a result of wrath of gods and therefore, could not be resisted by humans. Still, they could build their landmark structures to last, often by making them excessively stiff and strong, like the El Castillo pyramid at Chichen Itza.

Currently, there are several design philosophies in earthquake engineering, making use of experimental results, computer simulations and observations from past earthquakes to offer the required performance for the seismic threat at the site of interest.

These range from appropriately sizing the structure to be strong and ductile enough to survive the shaking with an acceptable damage, to equipping it with base isolation or using structural vibration control technologies to minimise any forces and deformations. While the former is the method typically applied in most earthquake-resistant structures, important facilities, landmarks and cultural heritage buildings use the more advanced (and expensive) techniques of isolation or control to survive strong shaking with minimal damage. Examples of such applications are the Cathedral of Our Lady of the Angels and the Acropolis Museum.

Trends and Projects

Some of the new state-of-the-art trends and/or projects in the field of earthquake engineering structures are presented below.

Proofed Low Cost Earthquake Building Material

This section may be written like an advertisement. Please help improve it by rewriting promotional content from a neutral point of view and removing any inappropriate external links. *(August 2010)*

A leading German composite construction company developed a proofed earthquake-safe supported core material RexWall - based on internal beam/frame constructions. The same principle allows construction of hurricane-safe houses.

Earthquake Shelter

One Japanese construction company has developed a six-foot cubical shelter, presented as an alternative to earthquake-proofing an entire building.

Concurrent Shake-table Testing

Concurrent shake-table testing of two or more building models is a vivid, persuasive and effective way to validate earthquake engineering solutions experimentally.

Thus, two wooden houses built before adoption of the 1981 Japanese Building Code were moved to E-Defence for testing. The left house was reinforced to enhance its seismic resistance, while the other one was not. These two models were set on E-Defence platform and tested simultaneously.

Combined Vibration Control Solution

Figure: *Seismically retrofitted Municipal Services Building in Glendale, CA*

Designed by architect Merrill W. Baird of Glendale, working in collaboration with A. C. Martin Architects of Los Angeles, the Municipal Services Building at 633 East Broadway, Glendale was completed in 1966. Prominently sited at the corner of East Broadway and Glendale Avenue, this civic building serves as a heraldic element of Glendale's civic center.

In October 2004 Architectural Resources Group (ARG) was contracted by Nabih Youssef & Associates, Structural Engineers, to provide services regarding a historic resource assessment of the building due to a proposed seismic retrofit.

In 2008, the Municipal Services Building of the City of Glendale, California was seismically retrofitted using an innovative combined vibration control solution: the existing elevated building foundation of the building was put on high damping rubber bearings.

Steel Plate Shear Walls System

Figure: *Coupled steel plate shear walls, Seattle, Copyright Mehdi Kharrazi*

A steel plate shear wall (SPSW) consists of steel infill plates bounded by a column-beam system. When such infill plates occupy each level within a framed bay of a structure, they constitute a SPSW system.

SPSW behaviour is analogous to a vertical plate girder cantilevered from its base. Similar to plate girders, the SPSW system optimizes component performance by taking advantage of the post-buckling behaviour of the steel infill panels.

The Ritz-Carlton/JW Marriott hotel building, a part of the LA Live development in Los Angeles, California, is the first building in Los Angeles that uses an advanced steel plate shear wall system to resist the lateral loads of strong earthquakes and winds.

Kashiwazaki-Kariwa Nuclear Power Plant is Partially Upgraded

The Kashiwazaki-Kariwa Nuclear Power Plant, the largest nuclear generating station in the world by net electrical power rating, happened to be near the epicenter of the strongest M_w 6.6 July 2007 Chûetsu offshore earthquake. This initiated an extended shutdown for structural inspection which indicated that a greater earthquake-proofing was needed before operation could be resumed.

On May 9, 2009, one unit (Unit 7) was restarted, after the seismic upgrades. The test run had to continue for 50 days. The plant had been completely shut down for almost 22 months following the earthquake.

Seismic Test of Seven-Story Building

Figure: *Seven-story wooden structure prior to shake-table testing, Japan*

A destructive earthquake struck a lone, wooden condominium in Japan. The experiment was webcast live on July 14, 2009 to yield insight on how to make wooden structures stronger and better able to withstand major earthquakes.

The Miki shake at the Hyogo Earthquake Engineering Research Center is the capstone experiment of the four-year NEESWood project, which receives its primary support from the U.S. National Science Foundation Network for Earthquake Engineering Simulation (NEES) Program.

"NEESWood aims to develop a new seismic design philosophy that will provide the necessary mechanisms to safely increase the height of wood-frame structures in active seismic zones of the United States, as well as mitigate earthquake damage to low-rise wood-frame structures," said Rosowsky, Department of Civil Engineering at Texas A&M University. This philosophy is based on the application of seismic damping systems for wooden buildings. The systems, which can be installed inside the walls of most wooden buildings, include strong metal frame, bracing and dampers filled with viscous fluid.

Seismic Retrofit

Seismic retrofitting is the modification of existing structures to make them more resistant to seismic activity, ground motion, or soil failure due to earthquakes. With better understanding of seismic demand on structures and with our recent experiences with large earthquakes near urban centres, the need of seismic retrofitting is well acknowledged.

Prior to the introduction of modern seismic codes in the late 1960s for developed countries (US, Japan etc.) and late 1970s for many other parts of the world (Turkey, China etc.), many structures were designed without adequate detailing and reinforcement for seismic protection. In view of the imminent problem, various research work has been carried out. Furthermore, state-of-the-art technical guidelines for seismic assessment, retrofit and rehabilitation have been published around the world - such as the ASCE-SEI 41, and the New Zealand Society for Earthquake Engineering (NZSEE)'s guidelines.

The retrofit techniques outlined here are also applicable for other natural hazards such as tropical cyclones, tornadoes, and severe winds from thunderstorms. Whilst current practice of seismic retrofitting is predominantly concerned with structural improvements to reduce the seismic hazard of using the structures, it is similarly essential to reduce the hazards and losses from non-structural elements. It is also important to keep in mind that there is no such thing as an earthquake-proof structure, although seismic performance can be greatly enhanced through proper initial design or subsequent modifications.

Strategies

Seismic retrofit (or rehabilitation) strategies have been developed in the past few decades following the introduction of new seismic provisions and the availability of advanced materials (e.g. fibre-reinforced polymers, FRP, fibre reinforced concrete and high strength steel). Retrofit strategies are different from retrofit techniques, where the former is the basic approach to achieve an overall retrofit performance objective, such as increasing strength, increasing deformability, reducing deformation demands while the latter is the technical methods to achieve that strategy, for example FRP jacketing.

- Increasing the global capacity (strengthening). This is typically done by the addition of cross braces or new structural walls.
- Reduction of the seismic demand by means of supplementary damping and/or use of base isolation systems.
- Increasing the local capacity of structural elements. This strategy recognises the inherent capacity within the existing structures, and therefore adopt a more cost-effective approach

to selectively upgrade local capacity (deformation/ductility, strength or stiffness) of individual structural components.

- Selective weakening retrofit. This is a counter intuitive strategy to change the inelastic mechanism of the structure, whilst recognising the inherent capacity of the structure.
- Allowing sliding connections such as passageway bridges to accommodate additional movement between seismically independent structures.

Performance Objectives

In the past, seismic retrofit was primarily applied to achieve public safety, with engineering solutions limited by economic and political considerations. However, with the development of Performance based earthquake engineering (PBEE), several levels of performance objectives are gradually recognised:

- Public safety only. The goal is to protect human life, ensuring that the structure will not collapse upon its occupants or passersby, and that the structure can be safely exited. Under severe seismic conditions the structure may be a total economic write-off, requiring tear-down and replacement.
- Structure survivability. The goal is that the structure, while remaining safe for exit, may require extensive repair (but not replacement) before it is generally useful or considered safe for occupation. This is typically the lowest level of retrofit applied to bridges.
- Structure functionality. Primary structure undamaged and the structure is undiminished in utility for its primary application. A high level of retrofit, this ensures that any required repairs are only "cosmetic" - for example, minor cracks in plaster, drywall and stucco. This is the minimum acceptable level of retrofit for hospitals.
- Structure unaffected. This level of retrofit is preferred for historic structures of high cultural significance.

Techniques

Common seismic retrofitting techniques fall into several categories:

One of many "earthquake bolts" found throughout period houses in the city of Charleston subsequent to the Charleston earthquake of 1886. They could be tightened and loosened to support the house without having to otherwise demolish the house due to instability.

The bolts were directly loosely connected to the supporting frame of the house.

External Post-tensioning

The use of external post-tensioning for new structural systems have been developed in the past decade. Under the PRESS (Precast Seismic Structural Systems), a large-scale U.S./Japan joint research program, unbonded post-tensioning high strength steel tendons have been used to achieve a moment-resisting system that has self-centering capacity. An extension of the same idea for seismic retrofitting has been experimentally tested for seismic retrofit of California bridges under a Caltrans research project and for seismic retrofit of non-ductile reinforced concrete frames. Pre-stressing can increase the capacity of structural elements such as beam, column and beam-column joints. It should be noted that external pre-stressing has been used for structural upgrade for gravity/live loading since 1970s.

Base Isolators

Base isolation is a collection of structural elements of a building that should substantially decouple the building's structure from the shaking ground thus protecting the building's integrity and enhancing its seismic performance. This earthquake engineering technology, which is a kind of seismic vibration control, can be applied both to a newly designed building and to seismic upgrading of existing structures. Normally, excavations are made around the building and the building is separated from the foundations. Steel or reinforced concrete beams replace the connections to the foundations, while under these, the isolating pads, or base isolators, replace the material removed. While the base isolation tends to restrict transmission of the ground motion to the building, it also keeps the building positioned properly over the foundation. Careful attention to detail is required where the building interfaces with the ground, especially at entrances, stairways and ramps, to ensure sufficient relative motion of those structural elements.

Supplementary Dampers

Supplementary dampers absorb the energy of motion and convert it to heat, thus "damping" resonant effects in structures that are rigidly attached to the ground. In addition to adding energy dissipation capacity to the structure, supplementary damping can reduce the displacement and acceleration demand within the structures. In some cases, the threat of damage does not come from the initial shock itself, but rather from the periodic resonant motion of the structure that repeated ground motion induces. In partical sense, supplementary dampers act similarly to Shock absorbers used in automotive suspensions.

Tuned Mass Dampers

Tuned mass dampers (TMD) employ movable weights on some sort of springs. These are typically employed to reduce wind sway in very tall, light buildings. Similar designs may be employed to impart earthquake resistance in eight to ten story buildings that are prone to destructive earthquake induced resonances.

Slosh Tank

A slosh tank is a large tank of fluid placed on an upper floor. During a seismic event, the fluid in this tank will slosh back and forth, but is directed by baffles - partitions that prevent the tank itself becoming resonant; through its mass the water may change or counter the resonant period of the building. Additional kinetic energy can be converted to heat by the baffles and is dissipated through the water - any temperature rise will be insignificant.

Active Control System

Very tall buildings ("skyscrapers"), when built using modern lightweight materials, might sway uncomfortably (but not dangerously) in certain wind conditions. A solution to this problem is to include at some upper story a large mass, constrained, but free to move within a limited range, and moving on some sort of bearing system such as an air cushion or hydraulic film. Hydraulic pistons, powered by electric pumps and accumulators, are actively driven to counter the wind forces and natural resonances.

These may also, if properly designed, be effective in controlling excessive motion - with or without applied power - in an earthquake. In general, though, modern steel frame high rise buildings are not as subject to dangerous motion as are medium rise (eight to ten story) buildings, as the resonant period of a tall and massive building is longer than the approximately one second shocks applied by an earthquake.

Adhoc Addition of Structural Support/Reinforcement

The most common form of seismic retrofit to lower buildings is adding strength to the existing structure to resist seismic forces. The strengthening may be limited to connections between existing building elements or it may involve adding primary resisting elements such as walls or frames, particularly in the lower stories.

Connections between Buildings and their Expansion Additions

Frequently, building additions will not be strongly connected to the existing structure, but simply placed adjacent to it, with only minor continuity in flooring, siding, and roofing. As a result, the addition may have a different resonant period than the original structure, and they may easily detach from one another. The relative motion will then cause the two parts to collide, causing severe structural damage.

Proper construction will tie the two building components rigidly together so that they behave as a single mass or employ dampers to expend the energy from relative motion, with appropriate allowance for this motion.

Exterior Reinforcement of Building

Exterior Concrete Columns

Historic buildings, made of unreinforced masonry, may have culturally important interior detailing or murals that should not be disturbed. In this case, the solution may be to add a number of steel, reinforced concrete, or poststressed concrete columns to the exterior. Careful attention must be paid to the connections with other members such as footings, top plates, and roof trusses.

Infill Shear Trusses

Shown here is an exterior shear reinforcement of a conventional reinforced concrete dormitory building. In this case, there was sufficient vertical strength in the building columns and sufficient shear strength in the lower stories that only limited shear reinforcement was required to make it earthquake resistant for this location near the Hayward fault.

Typical Retrofit Scenario & Solution

Soft-story Failure

This collapse mode is known as *soft story collapse.* In many buildings the ground level is designed for different uses than the upper levels. Low rise residential structures may be built over a parking garage which have large doors on one side. Hotels may have a tall ground floors to allow for a grand entrance or ballrooms. Office buildings may have stores in the ground floor which desire continuous windows for display.

Figure: *Partial failure due to inadequate shear structure at garage level. Damage in San Francisco due to the Loma Prieta event.*

Traditional seismic design assumes that the lower stories of a building are stronger than the upper stories and where this is not the case—if the lower story is less strong than the upper structure—the structure will not respond to earthquakes in the expected fashion.

Using modern design methods, it is possible to take a weak story into account. Several failures of this type in one large apartment complex caused most of the fatalities in the 1994 Northridge earthquake.

Typically, where this type of problem is found, the weak story is reinforced to make it stronger than the floors above by adding shear walls or moment frames. Moment frames consisting of inverted U bents are useful in preserving lower story garage access, while a lower cost solution may be to use shear walls or trusses in several locations, which partially reduce the usefulness for automobile parking but still allow the space to be used for other storage.

Beam-column Joint Connections

Beam-column joint connections are a common structural weakness in dealing with seismic retrofitting. Prior to the introduction of modern seismic codes in early 1970s, beam-column joints were typically non-engineered or designed. Laboratory testings have confirmed the seismic vulnerability of these poorly detailed and under-designed connections. Failure of beam-column joint connections can typically lead to catastrophic collapse of a frame-building, as often observed in recent earthquakes.

Figure: *Corner joint steel reinforcement and high tensile strength rods with grouted anti-burst jacket below*

For reinforced concrete beam-column joints - various retrofit solutions have been proposed and tested in the past 20 years. Philosophically, the various seismic retrofit strategies discussed above can be implemented for reinforced concrete joints. Concrete or steel jacketing have been a popular retrofit technique until the advent of composite materials such as Carbon fibre-reinforced polymer (FRP). Composite materials such as carbon FRP and aramic FRP have been extensively tested for use in seismic retrofit with some success. One novel technique includes the use of selective weakening of the beam and added external post-tensioning to the joint in order to achieve flexural hinging in the beam, which is more desirable in terms of seismic design.

Widespread weld failures at beam-column joints of low-to-medium rise steel buildings during the Northridge 1994 earthquake for example, have shown the structural deficiencies of these 'modern-designed' post-1970s welded moment-resisting connections. A subsequent SAC research project has documented, tested and proposed several retrofit solutions for these welded steel moment-resisting connections. Various retrofit solutions have been developed for these welded joints - such as a) weld strengthening and b) addition of steel haunch or 'dog-bone' shape flange.

Shear Failure within Floor Diaphragm

Floors in wooden buildings are usually constructed upon relatively deep spans of wood, called joists, covered with a diagonal wood planking or plywood to form a subfloor upon which the finish floor surface is laid.

In many structures these are all aligned in the same direction. To prevent the beams from tipping over onto their side, blocking is used at each end, and for additional stiffness, blocking or diagonal wood or metal bracing may be placed between beams at one or more points in their spans. At the outer edge it is typical to use a single depth of blocking and a perimeter beam overall.

If the blocking or nailing is inadequate, each beam can be laid flat by the shear forces applied to the building. In this position they lack most of their original strength and the structure may further collapse.

As part of a retrofit the blocking may be doubled, especially at the outer edges of the building. It may be appropriate to add additional nails between the sill plate of the perimeter wall erected upon the floor diaphragm, although this will require exposing the sill plate by removing interior plaster or exterior siding. As the sill plate may be quite old and dry and substantial nails must be used, it may be necessary to pre-drill a hole for the nail in the old wood to avoid splitting. When the wall is opened for this purpose it may also be appropriate to tie vertical wall elements into the foundation using specialty connectors and bolts glued with epoxy cement into holes drilled in the foundation.

Sliding Off Foundation and "Cripple Wall" Failure

Figure: *House slid off of foundation*

Single or two story wood-frame domestic structures built on a perimeter or slab foundation are relatively safe in an earthquake, but in many structures built before 1950 the sill plate that sits between the concrete foundation and the floor diaphragm (perimeter foundation) or studwall (slab foundation) may not be sufficiently bolted in. Additionally, older attachments (without substantial corrosion-proofing) may have corroded to a point of weakness. A sideways shock can slide the building entirely off of the foundations or slab.

Often such buildings, especially if constructed on a moderate slope, are erected on a platform connected to a perimeter foundation through low stud-walls called "cripple wall" or *pin-up*. This low wall structure itself may fail in shear or in its connections to itself at the corners, leading to the building moving diagonally and collapsing the low walls. The likelihood of failure of the pin-up can be reduced by ensuring that the corners are well reinforced in shear and that the shear panels are well connected to each other through the corner posts. This requires structural grade sheet plywood, often treated for rot resistance. This grade of plywood is made without interior unfilled knots and with more, thinner layers than common plywood. New buildings designed to resist earthquakes will typically use OSB (oriented strand board), sometimes with metal joins between panels, and with well attached stucco covering to enhance its performance. In many modern tract homes, especially those built upon expansive (clay) soil the building is constructed upon a single and relatively thick monolithic slab, kept in one piece by high tensile rods that are stressed after the slab has set. This poststressing places the concrete under compression - a condition under which it is extremely strong in bending and so will not crack under adverse soil conditions!

Multiple Piers in Shallow Pits

Some older low-cost structures are elevated on tapered concrete pylons set into shallow pits, a method frequently used to attach outdoor decks to existing buildings.

This is seen in conditions of damp soil, especially in tropical conditions, as it leaves a dry ventilated space under the house, and in far northern conditions of permafrost (frozen mud) as it keeps the building's warmth from destabilising the ground beneath.

During an earthquake, the pylons may tip, spilling the building to the ground. This can be overcome by using deep-bored holes to contain cast-in-place reinforced pylons, which are then secured to the floor panel at the corners of the building. Another technique is to add sufficient diagonal bracing or sections of concrete shear wall between pylons.

Reinforced Concrete Column Burst

Reinforced concrete columns typically contain large diameter vertical rebar (reinforcing bars) arranged in a ring, surrounded by lighter-gauge hoops of rebar. Upon analysis of failures due to earthquakes, it has been realised that the weakness was not in the vertical bars, but rather in inadequate strength and quantity of hoops. Once the integrity of the hoops is breached, the vertical rebar can flex outward, stressing the central column of concrete. The concrete then simply crumbles into small pieces, now unconstrained by the surrounding rebar. In new construction a greater amount of hoop-like structures are used.

One simple retrofit is to surround the column with a jacket of steel plates formed and welded into a single cylinder. The space between the jacket and the column is then filled with concrete, a process called grouting. Where soil or structure conditions require such additional modification, additional pilings may be driven near the column base and concrete pads linking the pilings to the pylon are fabricated at or below ground level. In the example shown not all columns needed to be modified to gain sufficient seismic resistance for the conditions expected. (This location is about a mile from the Hayward Fault Zone.)

Reinforced Concrete Wall Burst

Concrete walls are often used at the transition between elevated road fill and overpass structures. The wall is used both to retain the soil and so enable the use of a shorter span and also to transfer the weight of the span directly downward to footings in undisturbed soil. If these walls are inadequate they may crumble under the stress of an earthquake's induced ground motion.

One form of retrofit is to drill numerous holes into the surface of the wall, and secure short L-shaped sections of rebar to the surface of each hole with epoxy adhesive. Additional vertical and horizontal rebar is then secured to the new elements, a form is erected, and an additional layer of concrete is poured. This modification may be combined with additional footings in excavated trenches and additional support ledgers and tie-backs to retain the span on the bounding walls.

Brick Wall Resin and Glass Fibre Reinforcement

Brick building structures have been reinforced with coatings of glass fibre and appropriate resin (epoxy or polyester). In lower floors these may be applied over entire exposed surfaces, while in upper floors this may be confined to narrow areas around window and door openings. This application provides tensile strength that stiffens the wall against bending away from the side with the application. The efficient protection of an entire building requires extensive analysis and engineering to determine the appropriate locations to be treated.

Lift

Where moist or poorly consolidated alluvial soil interfaces in a "beach like" structure against underlying firm material, seismic waves travelling through the alluvium can be amplified, just as are water waves against a sloping beach. In these special conditions, vertical accelerations up to twice the force of gravity have been measured. If a building is not secured to a well-embedded foundation it is possible for the building to be thrust from (or with) its foundations into the air, usually with severe damage upon landing. Even if it is well-founded, higher portions such as upper stories or roof structures or attached structures such as canopies and porches may become detached from the primary structure. Good practices in modern, earthquake-resistant structures dictate that there be good vertical connections throughout every component of the building, from undisturbed or engineered earth to foundation to sill plate to vertical studs to plate cap through each floor and continuing to the roof structure. Above the foundation and sill plate the connections are typically made using steel strap or sheet stampings, nailed to wood members using special hardened high-shear strength nails, and heavy angle stampings secured with through bolts, using large washers to prevent pull-through. Where inadequate bolts are provided between the sill plates and a foundation in existing construction (or are not trusted due to possible corrosion), special clamp plates may be added, each of which is secured to the foundation using expansion bolts inserted into holes drilled in an exposed face of concrete. Other members must then be secured to the sill plates with additional fittings.

Soil

One of the most difficult retrofits is that required to prevent damage due to soil failure. Soil failure can occur on a slope, a slope failure or landslide, or in a flat area due to liquefaction of water-saturated sand and/or mud. Generally, deep pilings must be driven into stable soil (typically hard mud or sand) or to underlying bedrock or the slope must be stabilised. For buildings built atop previous landslides the practicality of retrofit may be limited by economic factors, as it is not practical to stabilise a large, deep landslide. The likelihood of landslide or soil failure may also depend upon seasonal factors, as the soil may be more stable at the beginning of a wet season than at the beginning of the dry season. Such a "two season" *Mediterranean climate* is seen throughout California.

In some cases, the best that can be done is to reduce the entrance of water runoff from higher, stable elevations by capturing and bypassing through channels or pipes, and to drain water infiltrated directly and from subsurface springs by inserting horizontal perforated tubes. There are numerous locations in California where extensive developments have been built atop archaic landslides, which have not moved in historic times but which (if both water-saturated and shaken by an earthquake) have a high probability of moving *en masse*, carrying entire sections of suburban development to new locations. While the most modern of house structures (well tied to monolithic concrete foundation slabs reinforced with post tensioning cables) may survive such movement largely intact, the building will no longer be in its proper location.

Utility Pipes and Cables: Risks

Natural gas and propane supply pipes to structures often prove especially dangerous during and after earthquakes. Should a building move from its foundation or fall due to cripple wall collapse, the ductile iron pipes transporting the gas within the structure may be broken, typically at the location of threaded joints. The gas may then still be provided to the pressure regulator from higher pressure lines and so continue to flow in substantial quantities; it may then be ignited by a nearby source such as a lit pilot light or arcing electrical connection.

There are two primary methods of automatically restraining the flow of gas after an earthquake, installed on the low pressure side of the regulator, and usually downstream of the gas meter.

- A caged metal ball may be arranged at the edge of an orifice. Upon seismic shock, the ball will roll into the orifice, sealing it to prevent gas flow. The ball may later be reset by the use of an external magnet. This device will respond only to ground motion.
- A flow-sensitive device may be used to close a valve if the flow of gas exceeds a set threshold (very much like an electrical circuit breaker). This device will operate independently of seismic motion, but will not respond to minor leaks which may be caused by an earthquake.

It appears that the most secure configuration would be to use one of each of these devices in series.

Tunnels

Unless the tunnel penetrates a fault likely to slip, the greatest danger to tunnels is a landslide blocking an entrance. Additional protection around the entrance may be applied to divert any falling material (similar as is done to divert snow avalanches) or the slope above the tunnel may be stabilised in some way. Where only small- to medium-sized rocks and boulders are expected to fall, the entire slope may be covered with wire mesh, pinned down to the slope with metal rods. This is also a common modification to highway cuts where appropriate conditions exist.

Underwater Tubes

The safety of underwater tubes is highly dependent upon the soil conditions through which the tunnel was constructed, the materials and reinforcements used, and the maximum predicted earthquake expected, and other factors, some of which may remain unknown under current knowledge.

BART Tube

For current BART information concerning various seismic retrofits see. A tube of particular structural, seismic, economic, and political interest is the BART (Bay Area Rapid Transit) trans-bay tube. This tube was constructed at the bottom of San Francisco Bay through an innovative process.

Rather than pushing a shield through the soft bay mud, the tube was constructed on land in sections. Each section consisted of two inner train tunnels of circular cross section, a central access tunnel of rectangular cross section, and an outer oval shell encompassing the three inner tubes.

The intervening space was filled with concrete. At the bottom of the bay a trench was excavated and a flat bed of crushed stone prepared to receive the tube sections. The sections were then floated into place and sunk, then joined with bolted connections to previously-placed sections. An overfill was then placed atop the tube to hold it down. Once completed from San Francisco to Oakland, the tracks and electrical components were installed. The predicted response of the tube during a major earthquake was likened to be as that of a string of (cooked) spaghetti in a bowl of gelatin dessert. To avoid overstressing the tube due to differential movements at each end, a sliding slip joint was included at the San Francisco terminus under the landmark Ferry Building.

The engineers of the construction consortium PBTB (Parsons Brinckerhoff-Tudor-Bechtel) used the best estimates of ground motion available at the time, now known to be insufficient given modern computational analysis methods and geotechnical knowledge. Unexpected settlement of the tube has reduced the amount of slip that can be accommodated without failure. These factors have resulted in the slip joint being designed too short to ensure survival of the tube under possible (perhaps even likely) large earthquakes in the region. To correct this deficiency the slip joint must be extended to allow for additional movement, a modification expected to be both expensive and technically and logistically difficult. Other retrofits to the BART tube include vibratory consolidation of the tube's overfill to avoid potential liquefying of the overfill, which has now been completed. (Should the overfill fail there is a danger of portions of the tube rising from the bottom, an event which could potentially cause failure of the section connections.)

Bridge Retrofit

Bridges have several failure modes.

Expansion Rockers

Many short bridge spans are statically anchored at one end and attached to rockers at the other. This rocker gives vertical and transverse support while allowing the bridge span to expand and contract with temperature changes. The change in the length of the span is accommodated over a gap in the roadway by comb-like expansion joints. During severe ground motion, the rockers may jump from their tracks or be moved beyond their design limits, causing the bridge to unship from its resting point and then either become misaligned or fail completely. Motion can be constrained by adding ductile or high-strength steel restraints that are friction-clamped to beams and designed to slide under extreme stress while still limiting the motion relative to the anchorage.

Deck Rigidity

Suspension bridges may respond to earthquakes with a side-to-side motion exceeding that which was designed for wind gust response. Such motion can cause fragmentation of the road surface, damage to bearings, and plastic deformation or breakage of components. Devices such as hydraulic dampers or clamped sliding connections and additional diagonal reinforcement may be added.

Lattice Girders, Beams, and Ties

Figure: *Obsolete riveted lattice members*

Lattice girders consist of two "I"-beams connected with a criss-cross lattice of flat strap or angle stock. These can be greatly strengthened by replacing the open lattice with plate members. This is usually done in concert with the replacement of hot rivets with bolts.

Figure: *Bolted plate lattice replacement, forming box members*

Hot Rivets

Many older structures were fabricated by inserting red-hot rivets into pre-drilled holes; the soft rivets are then peened using an air hammer on one side and a bucking bar (an inertial mass) on the head end. As these cool slowly, they are left in an annealed (soft) condition, while the plate, having been hot rolled and quenched during manufacture, remains relatively hard. Under extreme stress the hard plates can shear the soft rivets, resulting in failure of the joint.

The solution is to burn out each rivet with an oxygen torch. The hole is then prepared to a precise diameter with a reamer. A special *locator bolt*, consisting of a head, a shaft matching the reamed hole, and a threaded end is inserted and retained with a nut, then tightened with a wrench. As the bolt has been formed from an appropriate high-strength alloy and has also been heat-treated, it is not subject to either the plastic shear failure typical of hot rivets nor the brittle fracture of ordinary bolts. Any partial failure will be in the plastic flow of the metal secured by the bolt; with proper engineering any such failure should be non-catastrophic.

Fill and Overpass

Elevated roadways are typically built on sections of elevated earth fill connected with bridge-like segments, often supported with vertical columns. If the soil fails where a bridge terminates, the bridge may become disconnected from the rest of the roadway and break away. The retrofit for this is to add additional reinforcement to any supporting wall, or to add deep caissons adjacent to the edge at each end and connect them with a supporting beam under the bridge.

Another failure occurs when the fill at each end moves (through resonant effects) in bulk, in opposite directions. If there is an insufficient founding shelf for the overpass, then it may fall. Additional shelf and ductile stays may be added to attach the overpass to the footings at one or both ends. The stays, rather than being fixed to the beams, may instead be clamped to them. Under moderate loading, these keep the overpass centered in the gap so that it is less likely to slide off its founding shelf at one end. The ability for the fixed ends to slide, rather than break, will prevent the complete drop of the structure if it should fail to remain on the footings.

Viaducts

Large sections of roadway may consist entirely of viaduct, sections with no connection to the earth other than through vertical columns. When concrete columns are used, the detailing is critical. Typical failure may be in the toppling of a row of columns due either to soil connection failure or to insufficient cylindrical wrapping with rebar. Both failures were seen in the 1995 Great Hanshin earthquake in Kobe, Japan, where an entire viaduct, centrally supported by a single row of large columns, was laid down to one side. Such columns are reinforced by excavating to the foundation pad, driving additional pilings, and adding a new, larger pad, well connected with rebar alongside or into the column. A column with insufficient wrapping bar, which is prone to burst and then hinge at the bursting point, may be completely encased in a circular or elliptical jacket of welded steel sheet and grouted as described above.

Sometimes viaducts may fail in the connections between components. This was seen in the failure of the Cypress Freeway in Oakland, California, during the Loma Prieta earthquake. This viaduct was a two-level structure, and the upper portions of the columns were not well connected to the lower portions that supported the lower level; this caused the upper deck to collapse upon the lower deck. Weak connections such as these require additional external jacketing - either through external steel components or by a complete jacket of reinforced concrete, often using stub connections that are glued (using epoxy adhesive) into numerous drilled holes. These stubs are then connected to additional wrappings, external forms (which may be temporary or permanent) are erected, and additional concrete is poured into the space. Large connected structures similar to the Cypress Viaduct must also be properly analysed in their entirety using dynamic computer simulations.

Residential Retrofit

Wood Frame Structure

Predominantly residential/dwelling in North America consisted of wood-frame structure. Wood is one of the best materials for anti-seismic construction since it is of low mass and is relatively less brittle than masonry. It is easy to work with and very cheap compared to other modern material as steel and reinforced concrete. This is only resistant if the structure is properly connected to its foundation and has adequate shear resistance, in modern construction obtained by well connected surfacing of panels with plywood or oriented strand board in combination with exterior stucco. Steel strapping and sheet forms are also used to connect elements securely.

Retrofit methods in older woodframe structures may consist of the following, and other methods not described here.

- The lowest plate rails of walls are bolted to a continuous foundation, or held down with rigid metal clips bolted to the foundation.
- Selected vertical elements, especially at wall junctures and window and door openings are attached securely to the sill plate.

- In two story buildings using "western" style construction (walls are progressively erected upon the lower story's upper diaphragm, unlike "eastern" *balloon framing*), the upper walls are connected to the lower walls with tension elements. In some cases, connections may be extended vertically to include retention of certain roof elements.
- Low *cripple walls* are made shear resistant by adding plywood at the corners, and by securing corners from opening with metal strapping or fixtures.
- Vertical posts may be restrained from jumping off of their footings.

Wooden framing is efficient when combined with masonry, if the structure is properly designed. In Turkey, the traditional houses (bagdadi) are made with this technology. In El Salvador, wood and bamboo are used for residential construction.

Reinforced and Unreinforced Masonry

In many parts of developing countries such as Pakistan, Iran and China, unreinforced or in some cases reinforced masonry is the predominantly form of structures for rural residential and dwelling. Masonry was also a common construction form in the early part of the 20th century, which implies that a substantial number of these at-risk masonry structures would have significant heritage value. Masonry walls that are not reinforced are especially hazardous. Such structures may be more appropriate for replacement than retrofit, but if the walls are the principal load bearing elements in structures of modest size they may be appropriately reinforced. It is especially important that floor and ceiling beams be securely attached to the walls. Additional vertical supports in the form of steel or reinforced concrete may be added.

In the western United States, much of what is seen as masonry is actually brick or stone veneer. Current construction rules dictate the amount of *tie–back* required, which consist of metal straps secured to vertical structural elements. These straps extend into mortar courses, securing the veneer to the primary structure. Older structures may not secure this sufficiently for seismic safety. A weakly secured veneer in a house interior (sometimes used to face a fireplace from floor to ceiling) can be especially dangerous to occupants. Older masonry chimneys are also dangerous if they have substantial vertical extension above the roof. These are prone to breakage at the roofline and may fall into the house in a single large piece. For retrofit, additional supports may be added or it may be better to simply remove the extension and replace it with lighter materials, with special piping replacing the flue tile and a wood structure replacing the masonry. This may be matched against existing brickwork by using very thin veneer (similar to a tile, but with the appearance of a brick).

Seismic Response of Landfill

Solid waste landfills can be affected by seismic activity. The tension in a landfill liner rises significantly during an earthquake, and can lead to stretching or tearing of the material. The top of the landfill may crack, and methane collection systems can be moved relative to the cover..

Increasing the depth of a waste column typically leads to a decrease in the ground acceleration felt at the surface of that landfill during an earthquake.. The weight of waste is important in the analysis of landfill liner puncture and pipe crushing during an earthquake.. Unstable rock under landfills such as limestone may yield during an earthquake, leading to a partial collapse of the fill.. A major concern in this case would be the potential contamination of water sources that may be located below the landfill..

6

Earthquake-Resistant Construction

Field inspection and analyses of the performance of structures during earthquake shaking of their foundations have clearly shown that building design which blindly follows seismic code regulations does not guarantee safety against collapse or serious damage. The reasons are discussed in detail in and can be summarised as follows. First, there are large uncertainties in many of the aspects involved in the numerical design of structures, particularly in establishing the design earthquake shaking and in estimating the demands and predicting the supplies of the real three-dimensional soil-foundation-building (superstructure) system; second, the performance of the system depends on its state when the earthquake strikes - thus construction and maintenance, which includes repair, retrofitting and/or modifications, must also be considered in addition to the design aspects.

Importance of Construction and Maintenance

Design and construction of a structure are intimately related and the achievement of good workmanship depends, to a large degree, on the simplicity of detailing of the members and of their connections and supports. For example, in the case of a reinforced concrete structure, although it is possible to detail complex reinforcement on paper and even to realise it in laboratory specimens so that seismic behaviour is improved, in the field such design details may not be economically feasible. A design is only effective if it can be constructed and maintained.

Need For a Comprehensive Approach to Earthquake-resistant Construction

The need for a more comprehensive approach to the earthquake-resistant construction problem than that covered by existing seismic code procedures has been discussed by the author in several publications and has been summarised in. In a comprehensive approach to the design of a structure it is first necessary to establish the design criteria, that is, behaviour of the structure - serviceability, damageability, and safety against collapse. Once the design criteria are established, depending on the limit state controlling the design, the selection of the design earthquake(s) should be done according to the comprehensive approaches summarised in. In this comprehensive attempt to overcome the uncertainties involved in modelling the real three-dimensional soil-foundation-superstructure system and in the estimation of the demands and supplies, usually derived from numerical analysis, the design cannot be based on a single deterministic analysis of a single selected model. The designer should consider several models, based on possible ranges over which the parameters governing the behaviour of the real system can vary.

Importance of Conceptual Design

As discussed in several publications, the author believes that in order to overcome or decrease the uncertainties to which the values of most of the parameters in the estimation of the demands and supplies are subjected in any current seismic-resistant design procedure, it is necessary to pay more attention to conceptual design. Conceptual design is defined as the avoidance or minimisation of problems created by the effects of seismic excitation by applying an understanding of the behaviour rather than using numerical computations. Examples of conceptual design are given in. From the analysis of the basic design equations and the general equation for predicting response, it becomes clear that to overcome detrimental effects of the uncertainties in many of the factors in these equations the following philosophy can be applied: (1) control or decrease the demands as much as possible, and (2) be generous in the supply, particularly by providing large ductility with stable hysteretic behaviour (toughness).

Control or Decrease of Demands

Because of the uncertainties regarding the dynamic characteristics of future earthquake ground motions and their modifications as a result of the interaction of the soil with the foundation-superstructure system response, the conceptual idea would be to control the input to the structure foundation. One promising method is through the use of base isolation techniques including energy absorbing devices in the system. In the case of buildings, a decrease in demand can be achieved by a proper selection of the configuration of the building and its structural layout and by the proper proportioning and detailing of the structural and non-structural components, that is, by following the basic principles or guideline for achieving efficient seismic-resistant construction.

Ntegral Action of the Soil-foundation-superstructure System

As discussed previously, in order to obtain good performance of structures during severe seismic ground shaking it is necessary to analyse thoroughly the dynamic characteristics of the real three-dimensional soil-foundation (substructure)-superstructure system. Except for the cases where base isolation techniques and/or energy dissipation devices are used, the basic rule for earthquake-resistant design is to achieve integral action of each of the main parts of the system and between these main parts, that is, the whole substructure and superstructure should be tied together so that they can work as a unit.

Substructure

In earthquake-resistant foundation design, the following two main (basic) guidelines should be borne in mind: first, select a foundation layout and substructure system as simple as possible; and second, tie together the different elements of the substructure. The latter is of utmost importance in the case of a structure built at a site with poor, loose saturated granular soil, where moderate or strong ground-shaking (with effective peak accelerations exceeding about 0.15g) involving several cycles may cause permanent horizontal displacements due to lateral spreading and/or subsidence of the ground.

Pile Foundations

When the surface soils are very soft and/or can liquefy, piles can be used to advantage. However, they should be properly designed, keeping in mind the following considerations: first, as discussed above, the pile caps should be tied together with tie beams or a reinforced concrete slab that can work in tension and compression so that the foundation can act as a unit (assuring the integrity of the foundation); second, bearing, rather than friction piles, should be used if the foundation materials might liquefy; third, the piles should be able to carry not only axial but also shear and bending forces (which can be developed due to relative horizontal displacements between different layers in the soil deposit). Therefore, in the case of concrete piles, these should not only be longitudinally reinforced but also confined by suitable lateral reinforcement, particularly immediately below the pile cap illustrate that when properly designed piles are used, it is possible to design and construct buildings even in very soft soil that can liquefy.

Bridge Foundations & Supports

Dramatic collapse of bridges has been induced by failure of their foundations and/or supports and by the lack of integral action between the substructure and the superstructure. Liquefaction of loose saturated granular foundation soils has been a major source of bridge failure, of which perhaps the Showa Bridge collapse illustrated in has been the most publicized. Note that the collapse of the Showa Bridge was due to the relative movement of the bridge piers which were supported on steel piles driven through loose sands below the mud line. The distortion of the pile caused by the loss of lateral support from the liquefied sands induced the relative movement of the piers, causing the simple unconnected spans of the bridge to fall.

Analysis of the behaviour of numerous bridges with multi-simple spans supported on piers during the 1964 Alaska, 1964 Niigata, and the 1971 San Fernando Earthquakes showed clearly that they collapsed for lack of adequate connections at the supports. In view of this finding the State of California started a retrofitting program of similar existing bridges, tying the spans or segments of the superstructures by the use of restrainers.

Selection of Proper System and Configuration for the Superstructure

One of the most critical decisions influencing the ability of a superstructure to withstand earthquake ground shaking is the choice of its basic plan shape and configuration. The importance of a proper selection of the superstructure configuration will be discussed and illustrated for the case of building structures.

Building structures may be of many types and configurations and there is, of course, no universal ideal configuration for any particular type of building. However, there are certain basic or guiding principles of seismic-resistant design that can be used as guidelines in selecting an adequate building configuration structural layout, structural system, structural material and the non-structural components. These basic guidelines are as follows:

- Building (superstructure and non-structural components) should be light and avoid unnecessary masses

- Building and its superstructure should be simple, symmetric, and regular in plan and elevation to prevent significant torsional forces, avoiding large height-width ratio and large plan area
- Building and its superstructure should have a uniform and continuous distribution of mass, stiffness, strength and ductility, avoiding formation of soft stories
- Superstructure should have relatively shorter spans than non-seismic-resistant structure and avoid use of long cantilevers
- The non-structural components should either be well separated so that they will not interact with the rest of the structure, or they should be integrated with the structure. On the latter case, it is desirable that the structure should have sufficient lateral stiffness to avoid significant damage under minor and moderate earthquake shaking, and toughness with stable hysteric behaviour (that is, stability of strength, stiffness and deformability) under the repeated reversal of deformations which could be induced by severe earthquake ground motion. The stiffer the structure, the less sensitive it will be to the effects of the interacting non-structural components, and the tougher it is, the less sensitive it will be to effect of sudden failure of the interacting non-structural elements
- Superstructure should be detailed so that the inelastic deformations can be constrained (controlled) to develop in desired regions and according to a desirable hierarchy
- Superstructure should have the largest possible number of defence lines, that is, it should be composed of different tough structural subsystems which interact or are interconnected by very tough structural elements (structural fuses) whose inelastic behaviour would permit the whole structure to find its way out from a critical stage of dynamic response
- Superstructure should be provided with balanced stiffness and strength between its members, connections and supports
- The stiffness and strength of the entire building should be compatible with the stiffness and strength of the soil foundation.

It is of utmost importance that in seismic-resistant design the designer recognises from the beginning that the dynamic forces in a structure can be controlled by proper selection of its structural system and by the amount and distribution of its reactive masses (masses that will react to the shaking of the building foundation). The smaller the reactive masses, the smaller the earthquake forces (inertia forces). The use of unnecessary masses should be avoided; any mass used in the building should have a seismic-resistant function. Damage due to the use of unnecessary masses are illustrated infor nonengineered buildings and in for engineered buildings. Note the heavy earthfill (0.46 m) requiring a heavy slab in. This was located at a floor level where significant discontinuity in stiffness and strength existed in the lateral structural system. As a consequence of this unnecessary mass and of the structural.

Building and its Structure Should Be Simple, Symmetric and Regular in Plan and Elevation

Field inspections of earthquake performance of buildings demonstrate that the simpler the building the better the behaviour, all other parameters being similar. As pointed out by Dowrick , there are two main reasons for this: first, it is easier to understand the overall earthquake behaviour of a simple building than of a complex one; second, it is easier to understand, formulate in drawings, and construct simple structural details than complicated ones.

Symmetry and regularity in plan and elevation are desirable for much the same reasons. Symmetry is important in both directions of a plan. Lack of symmetry (in mass distribution and/or in stiffness, strength and ductility) leads to torsional effects which are difficult to assess properly and which can be very destructive. A plan layout with reentrant angles should be avoided A regular rectangular plan building with asymmetrical stiffening, as illustrated in should also be avoided because it will create significant torsional forces.

The reinforced concrete hotel in had a rectangular plan, but a very stiff service core area was placed anti-symmetrically near the far end of the building. The eccentricity induced by this service core was balanced in the upper stories (above the second story) by the solid masonry walls that were used as infills for the reinforced concrete frame located in the front end of this picture. However, these walls were interrupted in the second story (where the dining room was located) to have only glass so that there would be a view of the plants on the terrace and of the city.

Building and its Structure Should Have a Uniform and Continuous Distribution of Mass, Stiffness, Strength and Ductility

In shaking a building, an earthquake ground motion will search for every structural weakness. These weaknesses are usually created by sharp changes in stiffness, strength and/or ductility, and the effects of these weaknesses are accentuated by poor distribution of reactive masses. Severe structural damage suffered by several modern buildings during recent earthquakes illustrates the importance of avoiding sudden changes in lateral stiffness and strength.

A typical example of the detrimental effects that these discontinuities can induce is in the case of buildings with a 'soft story'. Inspection of earthquake damage as well as the results of analytical studies have shown that structural systems with a soft story can lead to serious problems during severe earthquake ground shaking.

The structure of the 10-story main building consists of a reinforced concrete frame with shear walls in the transverse (short) direction in stories 4 to 8. The shear walls were interrupted at the fourth floor level where there were 1.10-metre diametre reinforced concrete columns which were severely damaged. The exterior walls were all constructed of tiles.

This was the main damage to the building structure in and it occurred precisely at the discontinuity in the structural system (interruption of the reinforced concrete walls). There was also some damage to the exterior tile walls. This main building was retrofitted by continuing the reinforced concrete walls down to the foundation

Note the hinging at the top and bottom of the first story columns in. This first story was a 'soft story' because, except for glass all around, it was completely open, while the second story had walls and partitions that increased significantly the lateral stiffness of this second story relative to the first.

Note the large permanent lateral second floor level displacement of the main Treatment and Care Unit (a relative displacement with an interstory drift of 0.81 metres measured at the corner columns shown in This large interstory drift, which induced significant non-structural and structural damage and which led to the demolishing of the building, was a consequence of the formation of a soft story at the first story level because of the existence of a reinforced concrete wall above the second floor level.

Surface Faulting

The surface faulting illustrated in this slide occurred along the Beni Rached fault, a high angle normal fault that appears to be a secondary splay of the Oued Fodda fault which ruptured during the El Asnam Earthquake of October 10, 1980. The Richter magnitude of this event was The surface faulting occurred along a zone that extended at least 30 kilometres. In most places the surface faulting along the Beni Rached fault was expressed as a simple well-defined fault scarp with vertical displacement averaging about 1 metre. Numerous small graben and extension cracks occurred on the downthrown side along the base of the fault scarp. Apparent right and left lateral offsets of approximately 0.5 to 1 metre were measured along the fault. Many sheep and donkeys were trapped along the surface faulting.

Railroads

The east-west railroad between Algiers and Oran and passing through El Asnam crosses the main thrust of the Oued Fodda fault about 15 km east of El Asnam outside the town of Oued Fodda. At the time of the main shock a train going from El Asnam to Algiers was straddling the fault.

The train was completely overturned as illustrated in. It took about seven days before the work crews could clear the wreckage and reopen the railway. At another site near the location illustrated in this slide the railroad tracks were bent; it is suspected that this may have been caused by a secondary fault.

A Richter magnitude M = 7.9 earthquake struck the country of Guatemala on February 4, 1976 causing more than 22,000 deaths, 70,000 injuries and widespread devastation leaving over a million people homeless. The cause of this destructive earthquake was the rupture of the long Montagua fault which is the boundary between the Caribbean and North American plates. Surface faulting was seen for a distance of about 250 kilometres. The offset was left-lateral with an average slip of 1 metre and a maximum measured value of 3.25 metres. Ground rupture along the Montagua fault appeared as a series of parallel cracks.

Bridges

The only main highway between Guatemala City and Puerto Barrios (in the Gulf of Honduras) goes along the Montagua fault. At Agua Caliente, located about 27 kilometres from Guatemala City, there was a five-span steel plate girder bridge. The three central spans of this bridge were simply supported on four tall reinforced concrete piers (frames) as illustrated in. Because of the poor detailing of the supports for these spans (the short supported lengths) and the relative movement between piers, the three central spans collapsed shows the two main steel girders of one of these central spans, twisted by impact on striking the ground.

Lifelines

The M = 6.5 San Fernando Earthquake of February 9, 1971, which resulted in only 65 deaths but produced damage evaluated at more than $550,000,000 (1971 dollars), was caused by the rupture along the Santa Susana and the San Fernando faults. The fault offset (up to 1 metre vertically and 1 metre laterally) produced considerable damage to the lifelines in the San Fernando Valley. Illustrates the rupture of water and sewage lines that crossed the fault rupture. Some of the steel gas pipes were compressed and ruptured, causing gas leakage, resulting in fires.

Buildings

Damage to buildings and facilities along the fault scarps of the San Fernando Earthquake of February 9, 1971 varied considerably, ranging from completely demolished houses, to rupture of the foundations, tilting of the foundation slabs and walls (as illustrated in and to houses that suffered minor damage.

Internal Seismic Waves

During an earthquake, significant damage can result due to instability of the soil in the area affected by internal seismic waves. The soil response depends on the mechanical characteristics of the soil layers, the depth of the water table and the intensities and duration of the ground shaking. If the soil consists of deposits of loose granular materials it may be compacted by the ground vibrations induced by the earthquake, resulting in large settlement and differential settlements of the ground surface. This compaction of the soil may result in the development of excess hydrostatic pore water pressures of sufficient magnitude to cause liquefaction of the soil, resulting in settlement, tilting and rupture of structures.

The M = 8.6 Alaska Earthquake

This earthquake occurred in the Aleutian Alaska arc which is an important segment of the circum-Pacific belt of earthquakes where the Pacific Plate plunges downward and northward. About 6 percent of the large shallow earthquakes of the world occur in this region. During the 1964 Alaskan earthquake, the first slip occurred at a depth of approximately 30 kilometres under northern Prince William Sound, and the rupture of the rocks extended horizontally for 800 kilometres roughly parallel to the Aleutian trench. It has been estimated that about 200,000 square kilometres of the crust were deformed in this earthquake. It was the greatest area of vertical displacement ever measured in earthquake history. The vertical fault displacement in place on Montague Island amounted to 6 metres.

This earthquake which was recorded to have an intensity of M = 8.6 resulted in approximately 130 deaths and property damage estimated in 300 million dollars (1967 dollars). Of the 130 deaths, only 9 persons died from the effects of ground shaking: approximately 120 persons were drowned by very large 'tidal waves' (tsunami) produced by the sudden upward movement of the Alaskan sea floor along the rupturing fault. The greatest damage caused by the earthquake resulted from soil slides and from waves generated by those slides that occurred under water. The most spectacular landslide involving about 9.6 million cubic metres of soil took place at the Turnagain Height area of Anchorage, Alaska.

The slide extended approximately 1600 metres and extended inland an average of 280 metres. The Turnagain Height slide was a laterally spreading landslide caused by the combination of dynamic stresses and induced high pore water pressure in layers of soft clay and sands underlying the sliding mass. Within the slide area the original ground surface was completely devastated by displacements that broke up the ground into a complex system of ridges and depressions as illustrated in Slide J12. In the depressed areas, the ground dropped an average of 11 metres during the sliding. Houses in the area, some of which moved laterally as much as 150 or 180 metres.

The M = 7.5 Niigata, Japan, Earthquake

The Niigata Earthquake resulted in dramatic damage due to liquefaction of the sand deposits in the low-lying areas of Niigata City. In and around this city, the soils consist of recently reclaimed land and young sedimentary deposits having low density and shallow ground water table. At the time of this earthquake, there were approximately 1500 reinforced concrete buildings in Niigata City. About 310 of these buildings were damaged, of which approximately 200 settled or tilted rigidly without appreciable damage to the superstructure as is illustrated in. It should be noted that the damaged concrete buildings were built on very shallow foundations or friction piles in loose soil. Similar concrete buildings founded on piles bearing on firm strata at a depth of 20 metre did not suffer damage.

Civil engineering structures which were damaged by the Niigata Earthquake included port and harbour facilities, water supply systems, railroads, roads, bridges, airport, power facilities and agricultural facilities. The main reason for these failures was ground failure, particularly the liquefaction of the ground in Niigata City which was below sea level as a result of ground subsidence.

The M = 7.4 Caucete, Argentina, Earthquake

Approximately 65 persons were killed, 284 injured, and between 20,000 and 40,000 were left homeless by this earthquake. No surface faulting was detected, and the most notable effect of this earthquake was the extensive area of liquefaction (possibly thousands of square kilometres). Some of the effects of this liquefaction are illustrated in

The most dramatic effects of the liquefaction were observed in the town of Caucete, located approximately 70 kilometres from the epicentre. In this area not only were large amounts of sand boils observed but large linear and arcuate fissures were induced. Illustrates one of the fissures that appeared in a soccer field in Caucete after the soil was compacted and covered by more that 0.25 metres of water. In some places this fissure was 1 metre wide and more than 2 metres deep. Houses that straddled these fissures were damaged as illustrated in. In some of these houses, the ground floor slabs were covered by more than 10 centimetres of sandy mud.

The M = 7.7 Nihonkai-chubu, Japan, Earthquake

In spite of the fact that the epicentre of this earthquake was located approximately 100 kilometres offshore from Akita City, the seismic disturbance and the tsunami of this earthquake resulted in significant damage to coastal areas along the Japan Sea. The maximum ground acceleration recorded was 0.20g (200 gals) and it has been estimated that in one area the acceleration exceeded 0.30g (300 gals). However, the duration was about 60 seconds, about twice as long as those recorded previously. There were 104 deaths and 325 injuries. 5100 houses were either fully or partially destroyed, and significant damage was done to civil engineering structures, particularly port and harbour facilities. Most of the deaths (100 of the 104) were caused by the tsunami, and most of the damage to houses and structures was cause by liquefaction, as illustrated in illustrates the ground failure cause by liquefaction. The school that was built on this site and was supported on piles embedded down to a layer of soil offering large penetration resistance did not suffer structural damage in spite of the fact that as much as 0.5 metres subsidence developed at this site as a consequence of compaction as illustrated in shows the inside of the gymnasium (the basketball court) of this school that was intact in spite of the fact that the surrounding soil had subsided more than 0.30

metres. This illustrates and emphasizes the importance of proper foundations in earthquake-resistant design as will be discussed and illustrated later. One of the cranes (furthest from the camera) was derailed and its supports failed. Due to liquefaction of sand backfill the aprons (slab or pavement) cracked and collapsed.

Concluding Remarks

Avoiding serious damage is the main goal of earthquake-resistant construction. The seismic-resistant design provisions of most codes are concerned only with assuring an effective design and construction of structures against damage that might be induced by the vibratory response of the structure to the shaking introduced at their foundation by the ground. Illustrate the importance of damage caused by ground failure and the need for an analysis of the suitability of the site selected for the structure before its design and construction. While in certain cases of ground failure it is possible to design safe structures by proper design of their foundation, in other cases the only rational solution is a change of site. These are areas where the potential for severe ground failure is so high that government authorities should prohibit the building structures.

Ground Shaking

Integrated field inspection and post-earthquake analyses of structural damage that result from earthquake shaking is one of the most effective means of gaining knowledge on seismic response and improving the state of the art and of the practice in seismic-resistant design and construction. Such integrated inspection and analyses have revealed that besides the soil conditions discussed above, the seismic performance (degree of damage) of a structure is very sensitive to the following aspects (or perameters): type of foundation; configuration of the structure; structural material; and design and construction detailing.

The M = 6.7 Coalinga, California, Earthquake

A survey of single-family dwellings conducted in the City of Coalinga showed that nearly 300 houses, 15% of the total surveyed, were thrown off their foundations as illustrated in [14, 15]. Most of these dwellings were more than 20 years old and were supported on cripple studs without adequate lateral bracing, and several had been moved into the city from nearby oil fields and were placed directly on the smooth surface of concrete foundations without proper anchorage.

The M = 6.5 San Fernando, California, Earthquake

During this earthquake, structural sliding on foundations occurred not only in houses, but also in parts of some modern engineered buildings. This was the case of the reinforced concrete walls surrounding the service core of the five-story Medical Treatment and Care Unit of the Olive View Hospital. As illustrated in, these walls shifted on their supports at first floor level because they were purposely separated by a steel plate from the reinforced concrete supporting floor frame at first floor leve.

Structural Vibration

The seismic forces that develop during the vibratory response of a structure to earthquake ground shaking at its foundation are inertia forces whose intensity depends on the product of the mass and acceleration. Hence it is of the utmost importance to reduce the mass of the structure to a minimum. Thus when the designer is confronted with the problem of selecting the structural material, it is necessary to analyse the mechanical characteristics of available materials, normalized to their unit weight. In doing so, it becomes evident that among the traditional structural materials - timber, masonry, concrete and metals (steel and aluminium) - the most efficient earthquake-resistant material for low-rise buildings is timber. However, these buildings should be carefully designed and constructed, provided with proper lateral bracing and all of their components tied together from the roof down to the foundation. The importance of adequate lateral bracing and of tying together the components is illustrated in.

During the M = 6.5, February 9, 1971 San Fernando, California, Earthquake, numerous modern split level and two-story wood frame houses were severely damaged because at their first story level the garage walls were inadequately braced.

The effects of insufficient lateral bracing and of inadequately tying together all the components of a building were clearly demonstrated in the M = 6.7 May 2, 1983, Coalinga, California, Earthquake. Numerous porches collapsed (as illu) because they were not properly tied (anchored) to the main structure of the building and/or did not have their own proper lateral bracing.

Masonry Structures

Unreinforced masonry is very susceptible to damage during earthquake ground shaking. Solid brick masonry is very heavy and its tensile strength, and therefore its flexural strength per unit weight for in-plane and out-of-place seismic forces, is very small. Catastrophic failures of unreinforced masonry, as illustrated in, have occurred in many earthquakes.

Old unreinforced masonry buildings, whose walls are not properly connected to the floors, roof, and transverse walls (interior and exterior), constitute a threat to the occupants as well as to people that may be walking in the neighbourhood because the walls start to fall as soon as the building vibrates when subjected to even moderate ground shaking.

The devastating effect of even moderate earthquake shaking is illustrated in which shows what remained of the downtown area of the city of Managua after the M = 6.2, December 23, 1972 Managua, Nicaragua, Earthquake. A large number of the 5000 people killed in this earthquake were killed by falling masonry walls and roofs supported by these walls.

When ge panels of unreinforced masonry are used as veneer or infilling in wood, reinforced concrete or steel framed buildings, they also constitute a severe hazard because large portions of these panels can easily be dislodged from the frame.

Properly reinforced masonry can be used effectively in seismic-resistant construction, but unreinforced masonry can only be used in small panels that are properly framed (confined) by structural elements

Concrete Structures

Concrete is a relatively heavy material which, like masonry, has a very small (practically negligible) tensile, and thus flexural, strength. Therefore, it is usually reinforced with steel when used in structures. When the concrete is properly reinforced with steel it can be used effectively in seismic-resistant construction, but it still has relatively low strength per unit weight when normal weight aggregates are used. The use of lightweight aggregate concrete offers a significant advantage in seismic regions.

Steel Structures

Steel is a manufactured material, with usually excellent quality control, that is fabricated in structural shapes. While its stiffness per unit weight is practically the same as any other traditional constructional material, its strength and particularly its ductility and toughness per unit weight are significantly higher than concrete and masonry materials.

Because of its high strength per unit weight, the slenderness of steel structural members usually exceeds significantly the slenderness of similar structural members made of other traditional materials. Thus buckling becomes a serious problem, and the higher the yielding strength of the steel the greater the danger of buckling. Most structural shapes are formed by plate elements which can undergo local buckling, particularly when strained in the inelastic range.

Therefore, in earthquake-resistant design, the compactness requirements for the cross section of the critical regions of structural members are more stringent than for design against normal (standard) loading condition. Another problem in attaining efficient seismic-resistant construction of steel structures is in the field-connection of the structural members illustrate some of the above problems in attaining efficient earthquake-resistant design of steel structure. In during the 1971 San Fernando Earthquake the diagonal braces were stretched beyond their elastic limit.

The Cordova Building, shown in is a six-story office building with a penthouse. The earthquake-resisting structural system was designed to be a moment-resisting space frame (full moment resisting in the east-west direction and partial moment-resisting connections in the 20-cm reinforced concrete wall service core (near the north facade) where the penthouse was standing, and a 10-cm southeast corner, which contributed significantly in the seismic response of the building. The steel connections were shop welded and field bolted with high-strength bolts.

The main earthquake damage occurred in the first story and at the penthouse whose walls collapsed. The first and second story steel columns, located at the south end of the building, suffered severe local buckling as illustrated in As illustrated in the local buckling of the w14x30 southeast corner column, which occurred just below the second floor level, was so severe that the flanges tore away from the web and the web crimped, resulting in shortening of the column by about 3.8 centimetres. The mid-story stair landing was connected to this corner column, making it shorter and therefore stiffer than the other columns

7

Active Control Devices for Earthquake Resistance

After development of passive devices such as base isolation and TMD. The next logical steps is to control the action of these devices in an optimal manner by an external energy source the resulting system is known as active control device system. Active control has been very widely used in aerospace structures. In recent years significant progress has been made on the analytical side of active control for civil engineering structures. Also a few models explains as shown that there is great promise in the technology and that one may expect to see in the foreseeable future several dynamic "Dynamic Intelligent Buildings" the term itself seems to have been joined by the Kajima Corporation in Japan. In one of their pamphlet the concept of Active control had been explained in every simple manner and it is worth quoting here.

People standing in swaying train or bus try to maintain balance by unintentionally bracing their legs or by relaying on the mussels of their spine and stomach. By providing a similar function to a building it can dampen immensely the vibrations when confronted with an earthquake. This is the concept of Dynamic Intelligent Building (DIB).

The philosophy of the past conventional a seismic structure is to respond passively to an earthquake. In contrast in the DIB which we propose the building itself functions actively against earthquakes and attempts to control the vibrations. The sensor distributed inside and outside of the building transmits information to the computer installed in the building which can make analyses and judgment, and as if the buildings possess intelligence pertaining to the earthquake amends its own structural characteristics minutes by minute.

Control Force Devices

Many ways have been proposed to apply control forces to a structure. Some of these have been tested in laboratory on scaled down models. Some of the ideas have been put forward for applications of active forces are briefly described in the following:

Active-tuned Mass Dampers (TMD)

these are in passive mode have been used in a umber of structures as mentioned earlier. Hence active TMD is a natural extension. In this system 1% of the total building mass is directly excited by an actuator with no spring and dash pot. The system has been termed as Active Mass Driver (AMD). The experiments indicated that the building vibrations are reduced about 25% by the use of AMD.

Other Methods

The liquid sloshing during earthquakes has assumed significance importance in view of over flow of petroleum products from storage tank in post earthquakes. One of the important consideration with sloshing is that is associated with a very low damping. The wave height was controlled through force applied to the side wall by a hydraulic actuator. The active control successfully reduced wave heights to the level of 6% of those without control, for harmonic excitations at sloshing frequency. For earthquake type excitation the wave heights were reduced to 19% level.

Conventional approach to earthquake resistant design of buildings depends upon providing the building with strength, stiffness and inelastic deformation capacity. But the new techniques like Energy Dissipation and Active Control Devices are a lot more efficient and better.

Energy Dissipation Devices for Earthquake Resistant Building Design

Another approach for controlling seismic damage in buildings and improving their seismic performance is by installing Seismic Dampers in place of structural elements, such as diagonal braces. These dampers act like the hydraulic shock absorbers in cars – much of the sudden jerks are absorbed in the hydraulic fluids and only little is transmitted above to the chassis of the car. When seismic energy is transmitted through them, dampers absorb part of it, and thus damp the motion of the building.

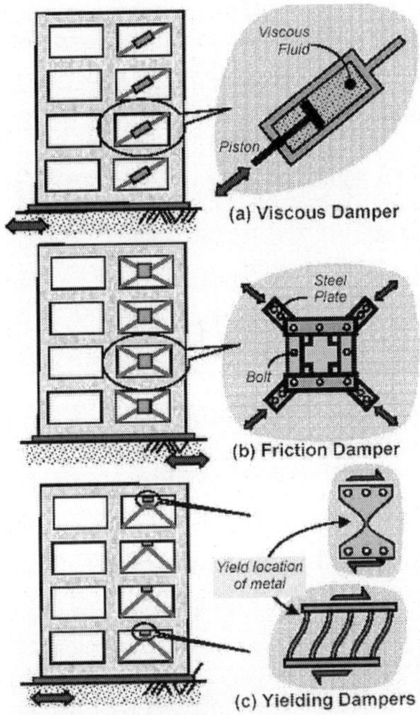

***Figure:** Energy Dissipation Devices*

Commonly Used Seismic Dampers

1. Viscous Dampers (energy is absorbed by silicone-based fluid passing between piston cylinder arrangement),
2. Friction Dampers (energy is absorbed by surfaces with friction between them rubbing against each other),
3. Yielding Dampers (energy is absorbed by metallic components that yield).
4. Viscoelastic Dampers (energy is absorbed by utilizing the controlled shearing of solids).

Thus by equipping a building with additional devices which have high damping capacity, we can greatly decrease the seismic energy entering the building.

How it Works?

The construction of a fluid damper is shown in. It consists of a stainless steel piston with bronze orifice head. It is filled with silicone oil. The piston head utilizes specially shaped passages which alter the flow of the damper fluid and thus alter the resistance characteristics of the damper. Fluid dampers may be designed to behave as a pure energy dissipater or a spring or as a combination of the two.

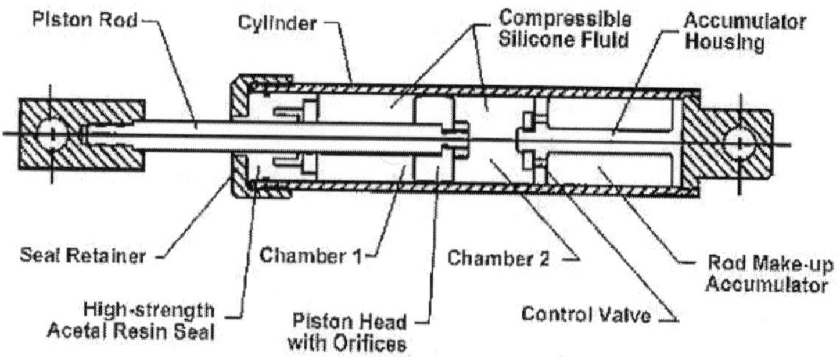

Figure: *How Dampers Work*

A fluid viscous damper resembles the common shock absorber such as those found in automobiles. The piston transmits energy entering the system to the fluid in the damper, causing it to move within the damper. The movement of the fluid within the damper fluid absorbs this kinetic energy by converting it into heat. In automobiles, this means that a shock received at the wheel is damped before it reaches the passengers compartment. In buildings this can mean that the building columns protected by dampers will undergo considerably less horizontal movement and damage during an earthquake.

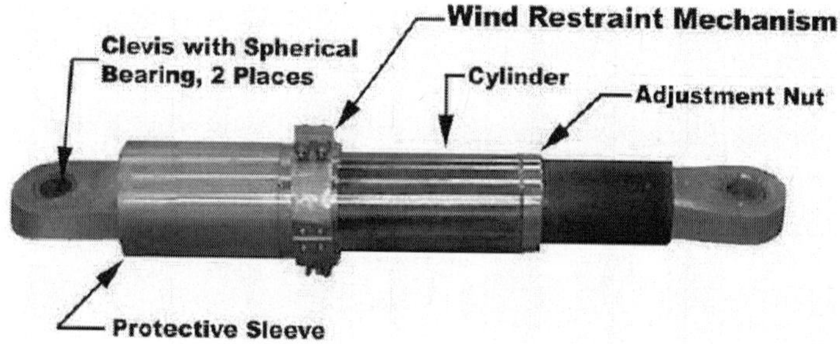

Figure: *Fluid Viscous Dampers*

New Breed of Energy Dissipation Devices

The innovative methods for control of seismic vibrations such as frictional and other types of damping devices are important integral part of seismic isolation systems as they severe as a barrier against the penetration of seismic energy into the structure. In this concept, the dampers suppress the response of the isolated building relative to its base. The novel friction damper device consists of three steel plates rotating against each other in opposite directions. The steel plates are separated by two shims of friction pad material producing friction with steel plates.

When an external force excites a frame structure the girder starts to displace horizontally due to this force. The damper will follow the motion and the central plate because of the tensile forces in the bracing elements. When the applied forces are reversed, the plates will rotate in opposite way. The damper dissipates energy by means of friction between the sliding surfaces.

The latest Friction-ViscoElastic Damper Device (F-VEDD) combines the advantages of pure frictional and viscoelastic mechanisms of energy dissipation. This new product consists of friction pads and viscoelastic polymer pads separated by steel plates. A prestressed bolt in combination with disk springs and hardened washers is used for maintaining the required clamping force on the interfaces as in original FDD concept.

Engineering Research for Non-Engineered Buildings

Studies of recent earthquakes have confirmed that loss of life occurs principally in single-family dwellings of unreinforced masonry, usually constructed by owners or local masons with site-found materials, such as ashlar or rubble stone, earth (adobe), or manufactured masonry block or brick. The 2005 earthquake in Pakistan resulted in more than 75,000 deaths from building collapses, primarily in rural dwellings and schools constructed mostly of stone and some manufactured masonry, some with a concrete bond beam and columns at corners, most with concrete roofs. The recent earthquake in Iran had similar results, but the structures there were constructed of earth rather than stone.

After disasters, many nations and organisations provide short-term and long-term relief. Donor organisations, such as the World Bank, which often provide or pay for replacement housing, want new structures to be earthquake-resistant both to ensure the safety of the occupants, and possibly to protect their investments. Plans for replacement buildings often call for concrete-masonry unit walls and concrete or wooden roofs. Although this kind of construction may be feasible in urban areas and towns and villages near roads, it may not be feasible in many other places. In the earthquake zone of Pakistan, for example, many villages are accessible only by trail. Thus construction materials for earthquake-resistant buildings would have to be carried long distances by hand or, at best, by mule. As a result, villagers often reconstruct or replace destroyed or damaged buildings using the same methods and materials that were used in the original construction.

One of the problems for people in remote, earthquake-prone areas is a lack of information about how to improve construction. In fact, most engineering research has been focused on the seismic rehabilitation of large, multistory structures rather than low-rise masonry and earthen buildings, and very little information is available on how to construct a simple building with built-in seismic safety.

Earthquake Response in the United States

Earthquakes in the United States have been followed by federally funded research to provide guidance to engineers on seismic strengthening. The loss of life in brick buildings in past earthquakes, particularly the 1971 Sylmar (Los Angeles area) earthquake, led to a research project funded by the National Science Foundation (NSF) carried out by a team with members from Agbabian Associates, S.B. Barnes and Associates, and Kariotis and Associates (ABK). Focused on unreinforced masonry buildings with wood-framed roofs and floors, the results of the project were reviewed by the professional community and later adopted into building codes. This earthquake-resistant construction, initially permitted by the city of Los Angeles as a "special procedure," has since gained acceptance and is now included in the Uniform Code for Building Conservation and the International Existing Building Code. The "special procedure" has been the basis for strengthening several thousand brick buildings. Although these provisions have not brought buildings up to current building code standards, they have reduced the chances of death and injury in earthquakes.

The 1994 Northridge (Los Angeles area) earthquake caused significant damage to steel-moment-frame building connections. In the aftermath, the Federal Emergency Management Agency (FEMA) funded research by a joint team of the Applied Technology Council (ATC), California Universities for Research in Earthquake Engineering (CUREE), and the Structural Engineers Association of California (SEAOC) to test and evaluate steel-moment connections This research led to a different detailing of steel joints for new buildings—several so-called FEMA connections and proprietary designs. Numerous other post-earthquake studies of concrete buildings have focused on beam-column joints and lightly reinforced buildings. In other parts of the world, however, much less has been done, especially for single-family dwellings and low-rise buildings.

Earthen Buildings

Earthen buildings constructed of a mixture of sand and silt with clay as the binder are found on all continents and in all countries. The most common types of earthen construction are adobe and rammed earth. Adobe bricks are made in a mold and are usually 16 to 20 inches long and 8 inches or more wide, a size that can be lifted by one person. Adobe buildings are constructed in a running-bond pattern with a mortar of adobe mud between blocks.

In the rammed-earth construction method, earth is packed into forms in a manner similar to the placement of concrete. The side of a formed unit may be as much as 4 feet high by about 6 feet long, depending on the thickness of the wall. Joints between units are packed with mud.

Historically, a bond beam, usually of wood, was used in earthen buildings. In the seismic zones of California, a concrete bond beam, or collar, is constructed at the top of walls, usually at the roof line; in some buildings, a parapet may be constructed above the bond beam. In recent years, engineers in California have attached the bond beam to the wall with vertical connector rods. However, in many places around the world, the bond beam is not connected to the wall at all.

Entire villages around the world are constructed using these methods. Some research has been done on the seismic behaviour of adobe construction in several countries, including Peru and the United States. The Getty Conservation Institute, through its Getty Seismic Adobe Program (GSAP), has supported testing of adobe construction and has published the results in several reports (Tolles et al., 2000).

Stone Buildings

In mountainous areas, stone has been the traditional construction material for walls. Stone walls are erected as typical masonry lay-up with bond blocks between wythes. In some cases buildings are constructed with single-wythe or unbonded, multi-wythe construction. The roof is constructed of wood trusses with a metal covering. Some later buildings were constructed with concrete bond beams and concrete corner columns. Many also had concrete roofs. A significant number of stone buildings collapsed in the Pakistan earthquake. Inspections after the earthquake revealed that the majority of collapsed buildings were the unbonded, single-wythe construction. These buildings did not have direct connections between bond beams and the stone walls, which might have reduced the number of collapses.

Brick Masonry Buildings

Brick construction, which is widely used in many countries, also has been the cause of many deaths and injuries in earthquakes. Research in the United States has led to many improvements; India has also developed strengthening procedures. In the United States, strengthening is based on the unreinforced wall acting as a vertical beam between the floor and the roof, or between floors in multistory structures. Connections to the roof and floors keep the walls in place.

In India, instead of a roof or floor diaphragm to brace the walls, the walls are allowed to span horizontally to perpendicular walls. The spacing between walls is limited and is within the traditional building wall-spacing, reflecting cultural preferences. In addition, seismic bands made of wire mesh plastered with a thin layer of concrete are placed at the roof, sill, and window lintel lines, and vertically at corners. This type of construction appears similar to the bond-beam approach, with additional ties at points where failures may occur. Interestingly, the placement of the seismic bands appears to be in line with research results on adobe buildings. It is not clear if this seismic-band type of construction is effective in all seismic zones, however.

Concrete Buildings with Masonry Infill

A common construction type used worldwide, especially for low-rise structures, is a concrete "frame" with unreinforced masonry infill. The "virtual diagonal strut" concept, in which the wall is regarded as a diagonal brace, is one way of evaluating such structures. Another is to consider the building a shear-wall structure. Out-of-plane loads require positive connections, usually epoxy-adhered bolts and metal connectors, between the wall and the bracing diaphragms.

Research Needs

It may not be possible to provide the levels of safety (life safety in the 475-year event and collapse prevention in the 2,500 year event) as envisioned in U.S. building codes for owner-built structures in other parts of the world. Nevertheless, all of the construction types outlined in this paper can be improved.

A number of studies and projects have been carried out over the years around the world, and many countries have assembled, or are assembling, building code provisions for different types of construction. However, these efforts have not been coordinated so that engineers and code authorities can make effective use of them. The Earthquake Engineering Research Institute online *World Housing Encyclopedia* may be a potential resource and repository for such information.

It has been suggested that a deterministic approach be taken in analysing simple, low-rise buildings. This would involve reviewing how these buildings fail and determining the sequence of failure. For example, we are aware that connections between elements are critical, and research might be directed toward improving connections between bond beams and walls. Another study might determine if a corrugated metal roof could be mobilised to act as a diaphragm with simple connections.

Another study might focus on improving single-wythe masonry with connections or stiffening elements to make structures safer. We also need guidelines for improving connections and the performance of low-rise, concrete frame buildings with masonry infill.

Improving existing owner-built buildings constructed with site-found materials would improve building performance and reduce the number of deaths and amount of damage in earthquakes. Cooperative efforts among nations could provide information to building owners and builders. Activities could be conducted by regional groups working with world bodies such as the United Nations or with individual countries.

A 21st Century Renaissance

Good afternoon, and congratulations to our new members. I know you will find your experience with the NAE rewarding. Few organisations have so great an effect on the future. And our influence is bound to grow. I come to that conclusion by analysing the Academy as if it were a business. Consider our customer base, which is the government and the public interest. What more influential constituencies could be served!? Consider our product, which is knowledge. It's growing in value as technology becomes more complex and plays a larger role in public policy. Beyond the strength of that business case, we have the good fortune to serve the NAE during an extraordinary period in history. As a nation, the United States has achieved a level of prosperity and technological leadership unequalled in human experience. No country has ever been so rich in resources. No country has ever been better positioned to improve the human condition through its technologies.

Of course there are obstacles. But there is one threat to progress we can eliminate, and that is low expectations! Given our current situation, we have good reason to set our sights high. We should expect advancements greater than those of the Italian Renaissance. I think it's both rational and helpful to envision a "21st Century Renaissance"-and picture the coming years as a period of unprecedented achievement.

The technology engine that drove the first Renaissance was the printing press. Today, it's digital computing and communication that allow faster, wider access to the best information, tools, and practices. Information technology is enabling faster development of more fundamental breakthroughs in virtually every field, including materials, energy, and biotechnology. If you measure progress in engineering by degrees of precision, we have come far. Engineers are now controlling photons of light with optoelectronics and working at the atomic level with nanotechnology. Few people would question our capacity for great technological achievements-certainly few in this audience.

The scale of accomplishments on the horizon makes the Renaissance idea rational. I believe that what makes it appealing is humanism, the force at the heart of the first Renaissance. It placed human needs and aspirations at the centre of every endeavor.

Assessing technology from a humanist perspective will, I think, be the greatest challenge we face at the NAE in the coming years. Technical knowledge is table stakes. We must also be aware of public priorities, environmental and health concerns, and difficult ethical issues such as privacy. We tend to think of these as "soft" issues. They defy the kind of certainty and mastery we bring to our technical disciplines. But these soft issues are the hardest to resolve.

Ultimate responsibility lies with policymakers. But we can help by considering these issues when we assess the impact of technology, including those elusive "unintended consequences." If we are to analyse costs and benefits, we need to understand what the public perceives as important and valuable.

Integrating human needs is engineering's biggest challenge and opportunity. The Internet exemplifies what can happen when we address, intentionally or inadvertently, those needs. It gives everyone what they wanted long ago: computers that talk to each other. What everyone initially received instead was cheap computing power. But people needed to communicate easily and inexpensively to use their computers productively.

Simplifying Complex Devices

I believe that one of the greatest opportunities in product engineering will be in simplifying complex devices. Computers and cameras and every digital device must and will become much easier for ordinary human beings to use. The end of the twentieth century will seem like the Dark Ages in terms of convenience. And engineering a new generation of digital products that are far more convenient will drive another cycle of economic growth.

As always, it is engineers who will develop these breakthroughs. In a sense, engineering is a humanist discipline by definition. Applying technology to human needs is what we are trained to do. Even many of the standards and measures we use in our work derive from the human form, beginning with the Egyptian cubit-from elbow to fingertips-and continuing through digits, palms, hands, feet, and yards.

It's interesting to trace engineering standards to their roots. Some of those standards are surprisingly arbitrary. There's a tale told at Kodak about the origin of the 35-mm film format, which was first used for motion pictures. It's said that George Eastman asked Thomas Edison how wide the film should be for Edison's new movie camera and viewer. Edison separated his thumb and forefinger and said: "About this wide."

Now, "about this wide" is hardly consistent with today's nanolevels of precision for international standards. It may be inconsistent with the truth, too. It's more likely that a roll of 70-mm film, which was common in the 1890s, was slit in half to make 35-mm film. If this is the true source of the standard, it's just as arbitrary as the legend. There's a similar story about the size of SRBs-solid rocket boosters-used to launch the first space shuttle. The SRBs had to be shipped by rail to the launch site. This influenced their design. Size was limited by the load capacity of railroad cars and the width of tunnels-all of which relate to the standard distance between rails, which is exactly 4 feet, 8-1/2 inches. That standard can be traced back to the first English tramways. They were built with the same jigs and tools used to build wagons. Spacing between wagon wheels had been standardised to accommodate ruts in long distance roads. Those ruts had been worn by wagons dating back to Imperial Rome, when chariots accompanied the legions that conquered Europe and Britain. Roman chariots were engineered with a very particular wheel spacing: just wide enough to accommodate the rear ends of two horses. So the next time you see a shuttle launch, remember that one of the major design features of one of the world's most advanced transportation systems was determined by the width of a horse's rear. And the next time you see a questionable design spec, it may be appropriate to ask: "What horse's rear end was responsible for this?" Joking aside, this is another reason the Renaissance metaphor is helpful, because it leads us to reconsider our most basic standards, assumptions, and priorities.

Engineering and Public Expectations

A Renaissance is a time of rethinking and optimism. Successful engineering has redefined the public's expectations of what we can accomplish. If we can put a man on the moon, why can't we significantly improve the environment and education? Why can't we eliminate hunger, poverty, and disease? In answering those questions, very few Americans would say we are limited by technology. For good reason, the general public has a positive view of technology. But they also see risks-to the environment, health, safety-and have other legitimate concerns. One of our roles at the NAE is to anticipate, to see those risks first, and to help policymakers deal with them effectively. It's an important contribution, because it helps sustain this country's confidence in change driven by technology. America's optimism and ability to change will put us at the centre of the 21st Century Renaissance. As will the strides made by American industry in being the first to bring new technologies to market. This confident, aggressive approach gives us a competitive advantage as a nation.

We are very good at competing in this country. Cooperation, on the other hand, is not so natural. Certainly not between government and industry. Not between regulators and the regulated. That must change to enable a Renaissance. Those of us in the business world need to relearn a lesson from the past: some government regulation is essential to capitalism. Half a millennium ago, capitalism grew from the bedrock of uniform legal and monetary systems. Today, the lack of a supportive legal system is a major source of Russia's difficulties. And Europeans are struggling to establish a uniform monetary system. Business leaders need to recognise and respect government's contribution to our success.

Many companies are benefiting from alliances and joint ventures with competitors. If competitors can cooperate creatively, so can business and government. We need to improve in this aspect, because the major projects that change life in the future will involve government, corporations, and universities. Government will often drive these projects, and it will rely on the NAE for guidance. The stakes in public policy decisions will be higher than ever before. All our wealth of resources and potential for historic achievements will make the risks bigger. This Academy was created to help policymakers avoid major pitfalls and stay on the path to progress. The NAE is a model of the Renaissance ideal, a forum for cooperation and sharing of knowledge from diverse institutions and specialities. To our work, we bring high standards of excellence and objective analysis. I'd like to propose that we also bring high expectations for the future. Let's think of our years here as the beginning of a 21st Century Renaissance-a time when technology, humanism, and cross-disciplinary cooperation combine to take us farther than we dreamed we could go. This is the picture we should keep in our minds. It's a vision that can help lead us to the great achievements we have every reason to expect in the years ahead.

Earthquake Resistant Buildings

Engineers would like to make every building earthquake-proof, but can't because it's too expensive. Instead, they recommend making dams and public buildings earthquake-proof. All other buildings should be earthquake resistant to avoid deaths. The cost of repair is a fraction of the cost of earthquake-proofing these buildings (Knowing Earthquakes, p. 2-3). The Transamerica Pyramid Building in San Francisco is earthquake-proof. It swayed more than one foot in the 1989 earthquake but wasn't damaged.

Procedure

A) A nine layer overlapped building with a cross brace inside was made on the earthquake machine. Straight buttresses and a roof attached with plasticine were added. The handles of the earthquake machine were pulled back to the line then let go. When the machine had stopped shaking, the handles were pulled again and again until the building collapsed. Each pull of the machine was timed to last two seconds. Three trials were done. At the end of each trial, the building was rebuilt. The data for each trial was recorded. Earthquake building with buttresses

B) A nine layer overlapped building with a cross brace inside was made and a roof was attached using plasticine. The building was built on a marble base isolator on the earthquake machine. The handles of the earthquake machine were pulled back to the line then let go. When the machine had stopped shaking, the handles were pulled again and again until the building collapsed. Each pull of the machine was timed to last two seconds. Three trials were done. At the end of each trial, the building was rebuilt. The data for each trial was recorded.

It is possible to make modifications to a building to make it earthquake resistant. The nine layer overlapped cubic building with buttresses, a roof, and cross braces was 83% more stable than the plain nine layer building. The nine layer overlapped cubic building with cross braces, a roof and a marble base isolator was 226% more stable then a plain nine layer building. The base isolator is more effective at stabilising the building than the buttresses because it actually absorbs more of the earthquake's energy.

Why Earthquake Resistant Structural Design

Buildings or structures are usually designed to bear only vertical loads. However, in the event of an earthquake, The ground shakes to & fro. This motion of the ground, imposes an additional horizontal load on the Building which it is unable to withstand unless specifically so designed; hence the damage.

Surprisingly, the additional cost to make a building/structure earthquake resistant is very low, usually about 5% of the structural cost of the building which comes out to be not more than 1-2% of the total cost of the building.

Reducing Earthquake Damage

Earthquakes are likely, knowing how to build can help reduce injury, loss of life, and property damage during an earthquake.

An earthquake-resistant building has a number of special structural features. Interior support walls called shear walls, made of reinforced concrete, strengthen the structure and help resist rocking forces. Shear walls in the centre of a building form a shear core. Cross-bracing reinforces walls with diagonal steel beams. Base isolators act as shock absorbers, and a moat allows a building to sway.

(World Book illustration by Doug DeWitt)

How to Build

A number of methods have been developed to build earthquake-resistant structures. These techniques range from extremely simple to fairly complex. For small- to medium-sized buildings, the simpler reinforcement techniques include making the structure ductile when subjected to earthquake. This ductility is achieved using properly placed reinforcement and providing support walls called shear walls. Shear walls, made of reinforced concrete (concrete with steel rods or bars embedded in it), help strengthen the structure and help resist rocking forces.

Sometimes, the buildings are protected using medium-sized devices that act like shock absorbers between the building and its foundation. These devices called base isolators, are usually bearings made of alternate layers of steel and an elastic material such as synthetic rubber. Also base isolators absorb some of the sideways motion that would otherwise damage a building

Virtual Research on Earthquake Resistant Structures

But this was no ordinary earthquake. In a groundbreaking series of tests, engineering researchers from UC San Diego's Jacobs School of Engineering jarred a full-size 275-ton building erected on a shake table, duplicating ground motions recorded during the January 17, 1994 Northridge earthquake in Los Angeles, California.

To record the impact on the building, the structure was fitted with some 600 sensors and filmed as the shake table simulated the earthquake, yielding a flood of data including stress, strain, and acceleration — so much information that engineers were having a hard time making sense of it all.

That's where visualization experts from the San Diego Supercomputer Centre (SDSC) at UC San Diego came in.

"By recreating the shake table experiment in movies in a virtual environment based on the observed data, this lets engineers explore all the way from viewing the 'big picture' of the entire building from a 360-degree viewpoint to zooming in close to see what happened to a specific support," said SDSC visualization scientist Amit Chourasia. "Integrating these disparate data elements into a visual model can lead to critical new insights."

Added José Restrepo, a professor of structural engineering at UCSD, "These visualizations give us an intuitive way to see how the building behaves in our shake table experiments — this tool will be very valuable in helping us understand the tests in ways we can't from traditional approaches, and also in sharing this research with other engineers and the public."

The costliest quake in U.S. history, the magnitude 6.7 Northridge event prompted calls for more scientific evaluation of structural elements, leading the engineers to conduct the building tests on one of the world's largest shake tables at UCSD's outdoor Englekirk Structural Engineering Centre.

A paper by Chourasia describing the project was published in a special graphics issue of ACM Crossroads, the student journal of the Association for Computing Machinery. In addition to helping engineers understand the earthquake's impact on the building, the visualizations can also give researchers a tool to do "what if" virtual experiments.

> *"We found that the recorded motion aligns very well with the movie we created," said Steve Cutchin, director of Visualization Services at SDSC. "This is important because knowing the model is accurate means it can be used to take simulated earthquake data and predict the sensor values — you can ask, 'What if a larger 7 point earthquake hits?' and simulate how the building will shake in response."*

To make the visualizations more useful and provide a rich visual context, the researchers wanted to incorporate recognisable elements from the surroundings, which meant integrating features from the actual video footage recorded during the test. "Our goal was to have fidelity not only in rendering quality but also in visual realism," said Chourasia. In addition, the integrated video would let the researchers validate this virtual reconstruction visually.

Once Chourasia and his colleagues had developed the building model and animated the deformation caused by the shaking, they worked to align the virtual camera and lighting with the real world video camera so that the scene would match in the recorded footage and the virtual version.

"However, when we tried to composite the actual video footage, we found that the instruments had sampled the data at 50 Hz but the video was recorded at 29.97 Hz.," Chourasia explained. "And there wasn't any timing synchronisation between the building sensors and camera." This posed a serious hurdle for compositing.

"After viewing the video footage, we noticed that the recording also contained audio data, because the moving building and shake table make noise, and this proved to be the key." By "listening to the building" and analysing the audio and sensor signals, the researchers were able to synchronise the video and instrument data for the visualization.

In the future, the visualization researchers plan to develop lighting models for more realistic rendering and to find automated ways to match the real and virtual cameras. They are also distilling lessons learned from this study into requirements for a visualization workbench for analysis of the dissimilar types of data that come from structural and seismic experiments.

In addition to providing visualization services for the shake table experiments, SDSC is also home to the NEESit Services Centre (NEESit) which is developing and maintaining a state-of-the-art grid to meet the cyberinfrastructure needs of the earthquake engineering community, including the UCSD facility and 14 other collaborating experimental sites across the nation.

Geological Survey Scientists, Research Help Haiti Reconstruction

USGS research will contribute to explicit recommendations to both the Haitian government and the international community that is assisting the reconstruction efforts," said Walter Mooney, USGS research geophysicist, who recently returned from Haiti.

The most recent USGS scientists travelling to Haiti are Carol Prentice and Rich Briggs, who arrived on Feb. 24, 2010. Prentice and Briggs will work with scientists from the University of Texas to measure coastal uplift. This USGS team of scientists is part of the Earthquake Disaster Assistance Team program, a new initiative between the USGS and the USAID Office of Foreign Disaster Assistance.

This research follows the work of Mooney and a team with expertise in seismology and earthquake engineering from the Earthquake Engineering Research Institute, who were in Haiti between Jan. 26 and Feb. 3, 2010. This team, supported by the United States Southern Command, surveyed Port-au-Prince to understand the geologic and engineering factors that contributed to the greatest damage from the shaking.

Mooney says new building codes for reconstruction in Haiti will be based, in part, on USGS research on geologic conditions that make some areas more at risk for damage than others. Currently, Haiti has no such standards in place, a factor that contributed to the recent widespread devastation.

> *"It's imperative that we move quickly," he adds, "because in some areas reconstruction has already begun that may not withstand another earthquake of this magnitude."*

Soil conditions, for example, play a big role in how a building fares during an earthquake. "Specifically, we found that buildings on harder, more stable bedrock fared much better than buildings on softer sediments, such as those located in the centre of cities like Port-au-Prince and Leogane," Mooney said.

The USGS scientists also installed seismic monitoring stations onto hard rock, as well as in the softer sedimentary basins. These monitoring stations precisely measure the location, frequency, and severity of the shaking, giving scientists the ability to assess the most dangerous and vulnerable areas.

Though an earthquake of this magnitude has not occurred since 1860, another large earthquake could strike Haiti in the near future. Beyond the immediate research following this earthquake, long-term monitoring using GPS will measure changes in the movement of the fault that runs through Haiti.

Earthquake Fatalities

In 2010, about 227,000 people were killed due to earthquakes, with over 222,570 from the magnitude-7.0 Haiti event, as reported by the United Nations Office for Coordination of Humanitarian Affairs.

According to official estimates, the Jan. 12 earthquake in Haiti injured 300,000 people, displaced 1.3 million, and left 97,294 houses destroyed and 188,383 damaged in Port-au-Prince and much of southern Haiti. This includes at least four people killed by a local tsunami in the Petit Paradis area near Léogâne. According to the USGS, this large, shallow earthquake produced violentshaking in Léogâne, west of Port-au-Prince. This shaking level can cause damage even to well-built buildings anywhere in the world. In Haiti, this high-intensity shaking together with both buildings vulnerable to earthquakes and high population exposure resulted in catastrophe. A magnitude-8.8 earthquake that hit offshore Bio-Bio, Chile, on Feb. 27 was the largest recorded in 2010. It killed at least 577 people, with about half of those deaths caused by an earthquake-generated tsunami. While the energy released by this earthquake was more than 500 times that of the one that struck Haiti, the fatalities were far fewer due to strict building codes in Chile and lower maximum shaking intensities.

The second-deadliest event in 2010 was a magnitude-6.9 earthquake that hit southern Qinghai, China, on April 14 (April 13, UTC time), leaving 2,968 dead. Overall, during 2010 earthquakes took the lives of people in 11 countries on four continents, including the countries of Afghanistan, Algeria, Argentina, Chile, China, Haiti, Indonesia, Iran, Mexico, Serbia and Turkey. The numbers for all events listed include those missing and presumed dead. Earthquakes injured people in nine additional countries, including Australia, New Zealand, India, Ethiopia, Peru, Solomon Islands, Taiwan, Venezuela and the United States, in California and Oklahoma.

As usual, the biggest earthquake in the United States in 2010 was in Alaska's Aleutian Islands. This year it was a magnitude-6.6

event on July 18 in the Fox Islands, and it caused no damage or casualties. The biggest 2010 earthquake in the contiguous United States was a magnitude 6.5 that shook Northern California on Jan. 10. About 30 people were injured and moderate damage occurred to hundreds of homes and buildings in the Eureka-Ferndale area. The event was felt as far away as Portland, Ore. A series of minor earthquakes peppered Oklahoma throughout the year, including a magnitude 4.4 on Oct. 13 that injured two people in Norman. An unusual earthquake occurred near the edge of the Continental Shelf about 125 km south-southeast of Westhampton, Long Island, New York, on Nov. 30. Because it was offshore, the magnitude-3.9 tremor caused no damage, but it was felt throughout Long Island, in large parts of Connecticut and New Jersey, and as far away as Maine and West Virginia. A magnitude-2.1 earthquake in New Jersey on Christmas Day and a magnitude-3.4 in Washington, DC, on July 16 round out the more unique seismic events recorded. The latter shook the windows in the White House, and the USGS received over 21,724 reports on the Did You Feel It? website .

The USGS estimates that several million earthquakes occur throughout the world each year, although most go undetected because they hit remote areas or have very small magnitudes. The USGS National Earthquake Information Centre publishes the locations for about 40 earthquakes per day, or about 14,500 annually, publishing worldwide earthquakes with a magnitude of 4.5 or greater or U.S. earthquakes of 2.5 or greater. On average, only 18 of these earthquakes have a magnitude of 7.0 or higher each year.

In 2010, 22 earthquakes reached a magnitude of 7.0 or higher, including the Chile quake that exceeded magnitude 8.0. These numbers are higher than those of 2009, which experienced 17 earthquakes over magnitude 7.0, including one over 8.0. While 22 earthquakes of magnitude 7.0 or greater is more than the average per year, and is the largest number of big events since 1968, it is still substantially fewer than 1943, which experienced 32 earthquakes of that size. Factors such as the size of an earthquake, the location and depth of the earthquake relative to population centres, and the fragility of buildings, utilities and roads all influence how earthquakes will affect nearby communities.

To monitor earthquakes worldwide, the USGS National Earthquake Information Centre receives data in real-time from nearly 990 stations in 85 countries, including the 150-station Global Seismographic Network, which is jointly supported by the USGS and the National Science Foundation and operated by the USGS in partnership with the Incorporated Research Institutions for Seismology (IRIS) consortium of universities.

In the United States, earthquakes pose significant risk to 75 million people in 39 states. The USGS and its partners in the multi-agency National Earthquake Hazard Reduction Program are working to improve earthquake monitoring and reporting capabilities via the USGS Advanced National Seismic System. Fibre reinforcing materials are used on inside and outside of various industrial materials, such as packaging, interior materials, concrete, and blades of wind-power generation. It increases the strength of the base material. We manufacture two fibre reinforcing materials, open mesh laid scrim fabric, "CRENETTE" and multi-axial warp knitted fabric "KURAMAS". These two materials are attracted in high-performance reinforcing effects.

Technical

The Hydraform building system replaces the conventional brick and mortar by using Hydraform blocks. The other components of the conventional building system remain largely unchanged. Hydraform system is a largely dry stacked-Interlocking masonry system that enables speedier construction of high quality, aesthetic and affordable building. The Blocks have an extremely appealing face-brick finish and provide a pre-pointed straight masonry.

The walls may be left exposed, plastered or finished with cement paint. These blocks are made using Hydraform Block making machine with an option to make at the site of construction itself. The interlocking construction is suitable for both Load bearing structures as well as framed structures – single or multistory. In conventional construction, brick could be good but is mortar used is normally weak thus making the over all masonry weak. This is taken care by Hydraform blocks which virtually use NO MORTAR making masonry uniform and strong.

The user also has option to use mortar slurry within the blocks. These are solid blocks for all kind of wall applications. Blocks can be made with fly ash based combinations or soil based combinations to meet relevent technical requirements.

Earthquake Resistant Construction

Hydraform interlocking block masonry can be effectively used in Earthquake resistant constructions incorporating necessary structural design and reinforcements – horizontal / vertical. HUDCO in consultations Dr. Arya, Seismic advisor to Gujarat State Disaster Management Authority arrived at suitable designs to use these blocks in Earthquake resistant houses and buildings in Gujarat.

HUDCO & other NGOs implemented these projects Hydraform machine made compressed earth interlocking block masonry. Vertical & Horizontal reinforcements can be effectively done for Earthquake construction. We continue to improve Earthquake construction techniques using Hydraform Building Systems with its continuous research process to improve the construction standards to meet Earthquake construction requirements.

Technical Papers on Hydraform Technology

Interlocking Dry Stacked masonry has been a topic of interest for mass usage and Research & Development among the technical and construction fraternity world wide. Its Hydraform who has pioneered in developing its interlocking building technology to mass scale application in various countries. Construction Projects, technical experts like architects / engineers / academicians, artisans, contractors are increasingly becoming cautious of advantages and economics of interlocking masonry. Hydraform gives guidelines for interlocking construction to be used in line with local building code.

Based on past experience with technology, Mr. Deepak Bansal (Structural Engineer), Asstt chief, HUDCO presented a technical paper on the application and technical features of Interlocking Dry Stacked Masonry on 8th International Masonry Conference 2010, Dresden, Germany. It may be noted that the views in the paper are of the presenter and HUDCO may or may not subscribe to the same.

Raw Material for Making Blocks

The blocks can be made in Hydraform machine using Fly ash – Lime – Gypsum – Cement- Sand, Fal-G or other fly Ash based tested combination OR Soil-Cement (compressed Stabilised Earth Blocks) OR Stone Dust - Cement. We can provide mix design assistance in this respect. Other option with Hydraform machine is to make Blocks using as well.

Raw Material Mix

Different mix designs are possible depending upon the quality of raw material and strength required, please consult our technical team in this respect.

Fly ash Cement Block

- Fly ash = upto 60%
- Coarse sand = upto 70%
- Stabiliser = 5% - 15% (including lime, cement, gypsum)

Stabilised Earth Block

- Soil = upto 70%
- Coarse sand = upto 30%
- Cement = 5 - 10%

Note: It is important to check the soil composition for contents like: Clay, Silt, fine and coarse sand. Normaly sandy - loam soils are better suited however other soils can be worked upon to get the right mix and strength. Black cotton soils are not suitable.

Block Strength: The block strength is determined by Soil / Fly Ash type, quantity of Cement used and other materials used and the extent of curing after manufacture of block.

We have obtained results as indicated below:

Soil - Cement Blocks		Fal-G - C. Sand	
Cement Content	**Strength**	**Cement Content**	**Strength**
5%	*4 to 5 MPa*	*5%*	*7 to 8 MPa*
10%	*7 to 8 MPa*	*10%*	*10 to 15 MPa*

1 MPa is equal to 10 Kg / Sq. cm

It is important to note that the results may vary and a proper training and technical assistance is must before you start the production and random sampling is must.

Thermal Quality:

Depending on ratio of soil or fly ash in mix, it can be up to 2-3 times better than conventional brick or concrete masonry.

Water Absorption:

Good quality blocks can be made from Hydraform machine to give:

For Fly Ash- Cement: Up to 15%.

For Stabilised Earth Block: Up to 12%.

Density

Density can vary with raw material used and its mix ratio. To suit respective application and technical perameters, blocks of different density can be made.

Fly Ash based Blocks can be made in range: 1500 to 1700 kg / cu.m

Stabilised Earth Blocks can be made in the range: +- 1950 kg / cu.m

Block Finish

Block finish is very good and compares even better than face brick finish. It provides option to keep the wall with or without plaster. In case of without plaster, the wall could be given a groove pointing / varnish / sili-coat or tereline coat.

8

Seismic Analysis of a High Rise Building

Professor Santiago Hernandez gave a special seminar at Ashurst Lodge describing the Seismic Analysis of a High Rise Building in Benidorm in which he described a special problem arising during the analysis of a large building on the Mediterranean Coast.

The structure was built in concrete rather than steel as is customary. It was also unusual in that there were no columns but only concrete shear walls which made the building unusually strong. The computational simulation was done using finite elements (66,518 elements and 181,828 nodes) resulting in more than 1 million degrees of freedom for the whole system after taking boundary conditions into consideration. It was analysed on a large PC using matrix partitions.

The dynamic analysis of the building produces a series of modes and frequencies. With these results, one can use the earthquake design spectrum defined by the Spanish codes. Due to the lack of earthquake data these codes have adapted a series of design spectra which are not too severe as the country has low seismicity.

Based on the seismic analysis as required by the codes, a new reinforcement scheme was proposed for the concrete structure. The building is now under construction and Santiago hopes to report on new developments next year when he plans to visit the Lodge again.

In the United States, the design of timber buildings is regulated by three documents: 1) the International Building Code (IBC) (2003), 2) the International Residential Code (IRC) (2003), and 3) the National Design Specification (NDS) (2005). The IBC and IRC provide the performance requirements for rationally designed and prescriptively designed buildings respectively, while the NDS provides the methods for designing the resistance of the building to meet the requirements of the IBC. This paper provides a brief discussion of how the methodologies used to design timber buildings in the United States has evolved to the point of sophistication it has today. Where the continued evolution is headed is covered with the directions of determining how the timber structural systems behave when mixed with other materials and how structural configurations and mixed structural systems will be some of the first issues to be dealt with in the near future. In addition, how test protocol effects results is being investigated, along with the specific issues of how the seismic response factors are to be determined for new systems are discussed.

In the United States, the design of timber buildings is regulated by three documents: 1) the International Building Code (IBC) (2003), 2) the International Residential Code (IRC) (2003), and 3) the National Design Specification (NDS) (2005). The IBC and IRC regulate the performance requirements for rationally designed and prescriptively designed buildings respectively, while the NDS provides the methods for designing the resistance of the building to meet the requirements of the IBC. The process for technically changing these documents is more or less an open process with all interested parties able to voice opinions and present technical information to the committees making decisions on whether proposed changes are in the public's interest or not. Earthquake engineering is a broad definition covering several sub disciplines. Mannvit's structural engineers have extensive knowledge and experience related to the different elements of earthquake engineering (also called seismic engineering), including seismic analysis and seismic design.

The primary objective of earthquake engineering is to make buildings that are earthquake resistant in order to save lives and property. Mannvit is based in Iceland, an active earthquake zone with peak ground acceleration (PGA) up to 0.4g and higher. Therefore, structures in Iceland are designed and constructed to withstand excessive seismic loads with minimal damage and therefore no loss of life. Mannvit's designers have experience in the seismic design of different structural systems of concrete, steel and timber, as well as non-structural elements such as warehouse racks and various building service systems.

One of the methods for connecting seismic forecasts to building design is through numeric modelling. Earthquake engineering is greatly enhanced by modelling the ways in which a certain type of structure will respond to seismic activity. Mannvit has experience in using the latest versions of advanced computer finite element analysis (FEA) programs such as SAP2000 and Etabs to assist in design and modelling. Depending on the form of the structures being considered, different analyses may be necessary, including linear and non-linear, as well as static, quasi-static and dynamic analysis. Mannvit´s projects include both residential and public buildings ranging from single family houses to complex buildings, such as power plants, dams, bridges, data centres.

Recent large projects include:

- The University of Reykjavik
- The Harpa Reykjavik Concert and Conference Centre

Our engineers have developed skills in their specialised disciplines in the constant presence of very active geology. In other words, our home and training ground is a relatively harsh environment, geologically, compared to many others. The skills of our engineers are based on developing safe, reliable structures and power projects in this environment. We have been well prepared to export these skills globally.

Laboratory Investigation of Building Structures

Seismic performance evaluation of building structures. Consulting services in design and construction of building structures.

Design

Design and Analysis of Earthquake Resistant Building Structures: Residental, Administrative, Public, Industrial Repair and Seismic Strengthening of Building Structures: Historic Buildings and Monuments Earthquake Protection of Building Structures and Historical Monuments using Innovative Techniques Adaptation, Reconstruction and Rehabilitation of Existing.

Experimental Testing

1. Testing of mechanical characteristics of materials, structural elements and assemblages
2. In situ forced and ambient vibration measurements of buildings, historical monuments and engineering structures
3. Quasi-static tests of joints, elements and assemblages
4. Shaking table tests of models of buildings, historical monuments and engineering structures to a reduced scale

5. Evaluation of seismic stability of historical monuments based on experimental testing and numerical modelling
6. Verification and upgrading of seismic strengthening systems based on experimental tests and numerical modelling
7. Machinery and foundation vibrations of turbo-generators at power plants
8. Dynamic testing of hydro, thermal and nuclear power plants facilities
9. Evaluation of earthquake floor response spectra based on "in situ" dynamic tests
10. Seismic effects of explosions and optimisation of mining procedure
11. Seismic qualification tests for mechanical and electro-mechanical equipment - "in situ" and in laboratory conditions
12. Mechanical tests of stockbridge and spacer dampers for high voltage power transmission lines
13. Seismic and vibration isolation of buildings and engineering structures

Professional and Consulting Services

- Revision of technical documentation for newly designed building structures
- Professional supervision during construction of building structures
- Professional supervision during repair and strengthening of building structures
- Final approval of works done on the newly constructed structures
- Consulting and supervision in the field of experimental mechanics
- Assistance (to Governements) in post-earthquake damage surveys and recovery

Innovations and Patents - Development and Verification

- Development of innovative technologies and structural systems in earthquake resistant design of structures
- Development of original methodologies for repair and strengthening of building structures
- Implementation of advanced materials in seismic upgrading of building structures

Software in Use

Development of new technologies and structural system Performance based design Composite and advanced engineering materials. Health monitoring of buildings, historical monuments and engineering structures Biomechanics and human reaction to vibrations Building structures under shock and impact Sustainability of building structures Detailed investigation of the principles, the criteria and the rules of the European Standard for Seismic Design or EN1998: Design of Structures for Earthquake Resistance", (EN 1998-1:2004).

Development of procedures for as much as possible realistic definition of bearing and deformability capacity of structural elements for the defined performance levels incorporated in EC8, in order to perform realistic evaluation of seismic resistance of RC buildings based on EC8.

Quantitative definition of engineering indicators as criteria for the performance of structural elements and integral structures.

Development of a handbook devoted to theoretical approaches used in EC-8 and numerical examples illustrating the application of the analytical methods.

Investor: European Commission

The main objective of the project is to study the seismic stability of historical buildings and develop specific strengthening techniques by using of advanced materials and technology that is reversible (possibility to be replaced by a new one). The consortium consists of fifteen universities from 12 countries of the European and the Mediterranean region. IZIIS participates in WG7 together with the Civil Engineering Faculty within the "Ss. Cyril and Methodius" university. The main activities of IZIIS are related to experimental testing of three models on the shaking table (model of a mosque, a cathedral and a church) as well as ambient vibration testing of prototype structures: Fossa Nova Abbey- Italy, Belerbey Palace, Istanbul, Mustafa Pasha mosque- Skopje and St. Nicholas church- Psacha, Macedonia. The in situ tests as well as the model test of the mosque were realised, while the models of the cathedral and the churches are under construction (November 2007)

The main objective of the project is to investigate the seismic stability of the brick-masonry walls at Beauharnois powerhouse in Quebec, Canada. The length of the powerhouse is one kilometre, and the height of the external wall is about 20m and it is made of brick masonry. The power plant is built in the first half of the past century. The experimental investigation was performed by in situ ambient vibration test of the wall and shaking table test of segments of the wall to a scale of 1/3 and 1/1. Experimental results have been used to perform refined non-linear numerical analysis of the walls applying the strengthening solution by implementation of anchor-dampers at critical points of the wall. The last phase of the project will be implementation of these anchor-dampers on the walls of the powerhouse.

The main objective of the project is to upgrade the seismic stability of this masonry structure by a DC-90 strengthening system. This system is developed by the DC-90 Company from Belgrade, Serbia. It consists of vertical, horizontal and diagonal steel ties for increasing the integrity of the masonry walls. All the works are performed from the outside, on the facades. Additionally, the dampers for dissipation of the seismic energy are implemented on diagonal braces. In order to design the required strengthening solution, in the first phase of the project, dynamic tests of the building have been performed to define the natural frequencies and stiffness characteristics of the building. After accomplishment of the strengthening, in the second phase, another dynamic test will be performed to check the effectiveness of the strengthening solution. IZIIS performed the ambient vibration tests in August 2007. After accomplishment of the strengthening works, the second test will be performed by forced vibrations to the structure simulating a moderate earthquake.

This project is related to seismic qualification of a high voltage circuit breaker according to IEC and IEEE international standards. The original sample is fixed to the shaking table and is tested separately in each of three orthogonal directions. The seismic input is usually sine-beat and/or some selected earthquake in real scale. The sample is instrumented by strain gages, LVDT's, accelerometres and linear potentiometres at different points to define the response of the test specimen in respect to acceleration, displacement and strain/stress. In this project, in the first phase, the circuit breaker type SDFL 11,12 single pole with six porcelains has been tested. In the second phase, the circuit breaker type SFEL 14, single pole with one porcelain, will be tested.

The project is being realised within the frames of the cooperation between IZIIS and the Civil Engineering Faculty at the University of Mostar. The bridge was completely destroyed during the war in1993 and rebuilt in 2003. The bridge is built by use of stone and represents a symbol of the City of Mostar.

The main objective of this on-going project is to evaluate the seismic stability of the bridge. In the first phase of the project, ambient vibration test has been performed in order to define the dynamic characteristics of the bridge: resonant frequencies, mode shapes and damping coefficients. In the second phase, numerical modelling based on experimental results will be performed in order to predict the seismic behaviour of the bridge.

This project is in its initial stage. The testing of the mechanical characteristics of the fibber concrete samples is related to a construction of a tunnel in Albania. The construction is under the supervision of Bechtel Company, USA. The first phase of the testing is related to testing of fibber concrete slab samples to bending, according to the international standards. It is a beginning of a long term cooperation between IZIIS and Bechtel that are to perform a variety of tests at the IZIIS laboratory to check the quality of the material used for construction of the tunnels and the bridges.

The main objective of this project is to evaluate the floor response spectra of the reactor building as well as the auxiliary buildings at the NPP Gentilly2, Quebec, Canada. The methodology represents a combination of experimental and numerical analysis which is new approach in the field of nuclear engineering. The new approach is related to experimental definition of transfer functions between particular floor and base level by performing ambient and/or forced vibration tests. The classical approach is based on analytical definition of transfer functions, which in some cases, could be non-realistic, if the mathematical model of the structure is not properly formulated. For this purpose, experimental measurements by special advanced equipment will be performed. The measurements will be performed in two parts: on the inside frame structure and on the outside containment. The testing started at the end of 2007.

Seismic Analysis

Multi-Canister Overpack (MCO) Handling Machine (MHM). The MHM handles large canisters of spent nuclear fuel in the Canister Storage Building at the Hanford Site. M&D performed a nonlinear time-history seismic analysis of the MHM. The analysis predicted that sliding and uplifting motion would occur during a seismic event. The results of M&D's analysis prompted the design of a new trolley uplift restraint for the MHM. M&D also performed in-situ testing of the

Electrical Control Cabinets for the MHM. The purpose of the testing was to determine the natural frequencies of the MHM trolley and the Electrical Control Panel Mounting Boards on the trolley operating floor. The results of the tests were used in computer models of the electrical panels to determine the In-Structure Response Spectra of the equipment for the purpose of seismic qualification of electrical relays and contactors.

Hanford K Basin Fuel Retrieval System (FRS) Seismic Analysis. M&D qualified the FRS safety class in-pool equipment for seismic loads using a nonlinear time-history analysis. The results of the analysis allowed the equipment to be qualified as free-standing and precluded the need for an expensive and complex seismic restraint system. A second part of the seismic analysis was to qualify the FRS equipment for accidental impact loads caused by falling equipment during a seismic event. M&D defined bounding accident conditions, demonstrated acceptable design margin by performing nonlinear impact analysis, and documented acceptable configuration of FRS components.

Sludge Transport System Trailer. M&D was responsible for demonstrating that the trailer would not overturn during a design basis seismic event. A detailed finite element model previously used for qualifying the trailer for transportation loads was modified to add the characteristics of the suspension system. Suspension air bags, roll stiffness, tire stiffness, and the landing gear interface with the ground were all simulated. Multiple time histories representing earthquake motions were applied to the model.

Seismic Qualification of PFP Storage Cubicles for New Fuel Storage Racks. The Plutonium Finishing Plant (PFP) stores Special Nuclear Material (SNM) contained within canisters. M&D prepared this report to qualify the design of the storage racks as a safety class (PC-3) component. M&D worked in partnership with the designer and fabricator (HiLine Engineering and Fabrication, Inc.), to ensure that the design met the appropriate structural criteria and functional requirements.

Fuel Transfer System. As part of the Hanford Spent Nuclear Fuel project, a new Fuel Transfer System has been added to the 50-year old K Basin concrete structure. The new system adds significant loads to the structure. M&D performed various analyses to qualify the structure for seismic and operational loads associated with the new equipment.

Structural Seismic Evaluation for the Cask Transport Trailer in the Canister Storage Building (CSB). M&D provided the analysis for this project to evaluate the Hanford Cask Transportation Trailer System for a postulated seismic event while parked in the CSB Facility. Trailer tipping and possible interaction hazards with adjacent safety systems were reviewed.

Computer Aided Design and Analysis of Reinforced Concrete Frame Buildings for Seismic Forces

For earthquake resistant design, evaluation of the seismic performance of buildings is essential to determine if an acceptable solution in terms of capacity and performance is achieved. Seismic performance evaluation requires static, modal and dynamic response analysis of a structure. The dynamic response history analysis often requires a large number of simulations using a number earthquake Ground Motion Records (GMR). Specialised computer programs for such analysis and design often lack the flexibility in user interaction as many of them come from research environments having focus only on the advanced analysis features. The study presents a user interface design and a scheme for automating the analysis process for evaluating the seismic performance of RC building frames. Such simulations involving a large number of earthquake GMR produce a huge amount of data that needs to be post processed in order to extract meaningful information about the behaviour of a building under earthquakes. The study also discusses the development of such post processor. The software tools developed here have been demonstrated through a case study, where a six storey RC frame building has been designed based on the seismic provisions of the current edition of the National Building Code of Canada and analysed to determine its seismic performance. The building is assumed to be located in Vancouver in western Canada. A set of eight simulated and sixteen actual ground motion records have been used in the inelastic dynamic time history analysis and a summary of the results have been presented.

Seismic loading provisions in most existing building codes focus on the minimum lateral seismic forces for which the building must be designed. Specifying the lateral forces alone is not enough to ensure that the desired level of performance will be achieved. A performance-based approach is desired in structural design where the structure is provided not only with adequate strength, but also appropriate deformability and damage tolerance for a given level of seismic hazard. Performance-based design (SEAOC, 1995) requires an accurate evaluation of performance of a structure at various stages in the design process and it requires reliable analysis of structures subjected to the design levels of loads. Although seismic design of buildings is performed based on the equivalent static loads method, current codes (NBCC, 2005) strongly recommend the use of dynamic analysis for the purpose of refinement in the design. Carrying out a detailed dynamic analysis of a structure using a number of earthquake Ground Motion Records (GMR) and constructing the performance profile of the structure in probabilistic terms, require enormous computing effort. Although the computing power of the modern computers is astounding, significant manual effort is needed in organising the input and output data from a given analysis tool and extracting meaningful information out of the huge quantity of analysis data. It is necessary to develop simple tools to automate such analysis and extract relevant information from a large set of analysis data and this study presents a general scheme for that with the implementation to a particular analysis software, namely IDARC2D (2006), a special purpose software tool for modelling the dynamic behaviour of Reinforced Concrete (RC) buildings subjected to earthquake ground motion.

Although there are a number of general purpose software packages available for structural analysis and design, special purpose software tools are often necessary to particular research needs. Since RC framed buildings are used in the study, IDARC2D is used here as it is a specialised software with the capability of analysing earthquake damage in multistory RC buildings and has a number of advanced analysis and modelling features which are not available in general purpose analysis software (IDARC2D, 2006). However, IDARC2D lacks a user interface and consequently, the input is written manually by the user to the text format and the output is also given in text form to be read by the user. This method of data input is cumbersome, especially if one plans to conduct dynamic analysis involving a large number of earthquake ground motion records and multiple building configurations and design choices. A set of interface tools have been developed in the present study to simplify the data input and interpretation of the analysis data. The tools presented here can be used for automating the input process and post processing the output data to conduct a large scale simulation of earthquake response of multi storey RC frame buildings.

To demonstrate the automation process developed here, a case study comprising a six storey RC frame building designed for Vancouver, Canada using the seismic provisions of the NBCC 2005 has been presented. The seismic performance of the building has been evaluated for a suite of eight simulated earthquake ground motion records compatible with the seismic hazard at Vancouver. To validate and compare the results obtained using IDRAC2D, the analyses have also been carried out using DRAIN2DX (Prakash *et al.*, 1993) a well known program for inelastic dynamic analysis of two-dimensional frames. However, DRAIN2DX is more general purpose seismic response analysis software as compared to IDARC2D. Modelling RC frames using DRAIN2DX is difficult and a simplified model using the two-dimensional beam-column elements has been used here.

Description of the Automation Tool for Analysis and Design

As discussed earlier, a typical automation scheme for structural analysis comprises the following three main components; pre-processor, analysis module and the post-processor, as shown in The pre-processor provides the user interface, preferably a graphical one through which the user can define the structural model, loads and support conditions. The pre-processing module then validates the models and conveys necessary information to the analysis module or engine. Once the analysis is completed, the output of that is processed by the post-processor which is designed to present the results in desired format. The post-processor may also be designed to process the output data from multiple sessions of analysis, which is typical the dynamic analysis using multiple GMR and compile the results to present the summary.

In this study, this scheme has been implemented using the computer program IDARC2D, which is an inelastic dynamic structural analysis tool used for detailed modelling of reinforced concrete building frames. One of the important features IDARC2D includes is automatic generation of the moment-curvature (M-Ö) relation for beam and the axial force-moment (P-M) interaction curves for columns given the section size and reinforcement details. It also has the option of specifying the hysteretic behaviour of the different elements of the structure. The use of infill panels, transverse beams, shear walls, different brace types is also another example of the many options available in this software. The user can choose to perform time-history, push-over or quasi-static analysis on the structure.

IDARC2D pre-processor: The pre-processor unit has been built using the Excel (Microsoft Corporation) and Visual Basic scripts. All input data are gathered in a Excel Workbook which provides appropriate forms with appropriate data labels to fill out with necessary data. The user can edit and change the editable fields and finally when the complete set of data for structural model definition, material properties, analysis options etc. are entered into the preprocessor, the user can instruct the pre-processor to prepare the IDRAC2D input files in text format by clicking on the "Write Input" button from the user interface After the input file is written, the user may run the IDARC2D program to generate the output files. The excel file contains seven worksheets where all data should be filled.

IDARC2D post-processor: Dealing with output files on the other hand requires reading a lot of data and generating graphs for visualising the response of the structure. This is done here using a graphical user interface developed in MATLAB as shown The post-processing interface developed in MATLAB scans through the output files generated by IDARC2D and provides tools for plotting a number of response quantities. The program can plot push-over curves, mode shapes, inter-storey drifts and time-history graphs

It also organises the output from multiple sessions of dynamic analysis with a number of earthquake GMR into summary tables or graphs as shown in (top right corner). The data processed by the post-processor can also be presented in the Excel format.

Shows the schematic architecture of the pre and post processing units as described earlier. The pre-processor engine is based on visual basic scripts or macros for manipulating an Excel workbook that gathers the input data necessary for IDARC2D. The input form and data cells in Excel are dynamically organised based on the type of analysis or the problem size. Once the data is gathered through the preprocessor, the user can instruct it to make appropriate data files in text format for IDARC2D. IDARC2D produces a number of output text files with general information and specific structural response, such as, storey drift, storey hysteris etc. The MATLAB based post-processor scan through the IDARC2D output files and produces necessary graphical output in order to visualize the analysis results, such as mode shapes, push-over curve, time-history of displacement or drift etc.

Generating beam-column properties: Two Excel files are created specifically for designing and calculating the capacity of rectangular reinforced concrete sections subjected to simple bending. Another Excel file is designed with the aid of VB scripts to generate the Axial Load-Bending Moment (P-M) interaction curves for rectangular reinforced concrete sections with uniform reinforcement. The input to these programs are the section size, reinforcement details, materials strength and resistance factors as defined in NBCC (2005). It must be noted that these tools are not required for IDARC2D analysis as the information produced by them can be internally generated by IDARC2D based on the section size, material properties and the reinforcement details. These tools are meant to be used with DRAIN2DX which does not have the capability to generate them internally.

Shows a screens hot of the Excel worksheet to generate the interaction diagram. The advantage of this worksheet is that the user has the freedom of choosing the strength reduction factors for concrete and steel. The modulus of elasticity of concrete and steel are also specified by the user. The sectional dimensions and reinforcement are among the input parameters as well as position of the reinforcing steel. The stress-strain distribution of concrete is represented by an equivalent stress block and the crushing strain of the concrete is specified by the user. After all the data are entered, the P-M interaction curve is generated automatically by pressing the Draw Diagram button. First, the output data is written in another worksheet labelled Output Data which is linked to the interaction diagram.

Case Study-a Six Storey Rc Frame Building in Vancouver

The use of the pre and post processor has been demonstrated in the following example. A six storey building in Vancouver has been designed using the NBCC (2005) seismic provisions. The geometric details of the building are shown in The building has several 6 m bays in the N-S direction and 3 bays in E-W direction. The E-W bays consist of two nine-metre office bays and a central 6 m corridor bay. The storey height is 4.85 m for the first storey and 3.65 m for all other storeys. The building is composed of a set of parallel frames equally spaced 6 m apart. A typical intermediate frame as shown in has been considered for the detailed design and analysis. The chosen cross sections of the columns and primary beams resulting from the design are shown in Table 1. All secondary beams are assumed to be of size 300x500 and the slab thickness is 120 mm.

Seismic provisions of NBCC 2005: The 2005 edition of NBCC allows the use of the equivalent static load method in the structural design against earthquake excitations. The seismic hazard is expressed in terms of a Uniform Hazard Spectrum (UHS), which provides the maximum expected spectral acceleration S_a of a Single-Degree-Of-Freedom (SDOF) system with 5% damping. The elastic base shear, V_e for a building with the single-degree-of-freedom system can be obtained by multiplying the spectral acceleration S(T) corresponding to the fundamental period of the building, T_a with the weight of the building, W. Considering the ductility capacity, the over-strength, the higher mode effects and the importance of the structure the design base shear V is given by Eq. 1:

$$V = \frac{S(T_a)M_v I_o W}{R_o R_d} \geq \frac{S(2.0)M_v I_o W}{R_o R_d} \quad (1)$$

where:

M_v = Accounts for higher mode effect,

I_e = The importance factor

R_d and R_o = Account for ductility and overstrength, respectively.

The design base shear is distributed along the height of the building according to provisions NBCC (2005).

Design of the building frame based on NBCC 2005: The building has been designed to resist the effect of the equivalent lateral loads combined with gravity loads; dead and live. The elements of the structures are designed based on the most critical load combination. The following load combinations have been used in the design: (a) the lateral load combination (D + 0.5L + E) and (b) the gravity load combination (1.25D+1.5L), where D is the dead load, L is the live load and E is the equivalent static earthquake force. The design base shear based on the NBCC 2005 provisions is 424 kN and the ductility and overstrength factors are 4 and 1.7, respectively.

The yield stress, f_y for reinforcing steel and the 28 day concrete compressive stress, $f2_c$ are assumed to be 400 and 30 MPa, respectively. Live load on the roof is assumed to be 2.2 kN m^{-2} on other floors it is 4.8 kN m^{-2} on the corridor bay and 2.4 kN m^2 on the other bays.

The static design however involves a few iterations until a safe and economic cross section is reached for all elements. Since the calculation of the base shear according to the NBCC (2005) requires the fundamental period of the structure, which is calculated using an empirical formula, the fundamental period should be checked against the one obtained from the modal analysis after the first design iteration. Usually the modal analysis of the bare frame structure gives a longer period for the fundamental mode of vibration as compared to the period computed using the empirical formula suggested in the code. If the fundamental period obtained from modal analysis is greater than the one obtained from the empirical formula, according to NBCC (2005), the design base shear needs to be revised to achieve a more realistic design load. The revision of the base shear should be based on a period which is 50% higher than that obtained from the empirical formula of NBCC (2005) or the one obtained from the modal analysis, whichever is less. In this case, the code defined formula ($T = 0.075(h_n)^{3/4}$) gives a period of 0.78 sec, while the modal analysis gives a value of 1.68 sec, which is more that 1.5 times the code defined value (1.17 sec). Thus the building needs to be redesigned for the base shear calculated using a period of 1.17 sec. The reinforcement details of the primary beams and columns are given in. First four mode shapes of the building are shown in and the corresponding periods are 1.68, 0.54, 0.3 and 0.2 sec.

Push-over analysis: A force and displacement controlled push-over analysis is performed to simulate the structure's response to incremental lateral loading. The push-over analysis serves as an important tool for estimating the strength and ductility capacities of the structure. The push-over curve obtained using the IDARC2D and DRAIN2DX are shown in

In the push-over curves as shown in, the base shear coefficient (V/W) is defined as the ratio of the base shear, V to the total tributory weight, W for the building frame. In this case, the design base shear coefficient is equal to 0.0733. The resulting push-over curves show that first occurrence of hinge formation in the frame element corresponds to a base shear coefficient of approximately 0.1 when IDARC2D is used and 0.09 when DRAIN2DX is used.

It's clear from that both programs give similar initial response, however there is a difference between the results in the post-yielding zone. The point where the maximum inter-storey drift reaches 2.5% (in this case this occurs at the first storey level) is marked on both the curves.

This point indicates the inter-storey drift limit in NBCC 2005 for collapse prevention performance. The maximum value of the inter-storey drift of 2.5% corresponds to a base shear coefficient of 0.139 and a roof drift (overall deformation) of 1.33% when IDARC2D is used and the corresponding base shear coefficient is 0.123 and the roof drift is 1.26% when DRAIN2DX is used.

Dynamic Analysis: The roof drift and inter-storey drift are important parameters to describe the overall deformation and performance of the structure. The roof drift or the total drift is the roof displacement expressed as a percentage of the total height of the building, while the interstorey drift is the differential displacement between current level to the one immediately below and expressed as the percentage of storey height. A dynamic analysis is performed using eight artificial ground motion records compatible with the seismic hazard at Vancouver (Tremblay and Atkinson, 2001).

Four of those records of long duration, while the other four records are of short duration. The properties of these ground motions are shown in The maximum inter-storey drifts of all floors along with the envelope and mean values have been compiled and plotted for all eight records (The maximum inter-storey drifts also occur at the first storey level and are equal to 2.07 and 1.46% due to the long duration and short duration records, respectively. The time-history of the first storey produced by the ground excitation, S1 is also plotted (to show the displacement of this storey during the earthquake period and a few seconds later. The total duration

of the earthquake is 8.53 sec. It is clear from the time-history graph that the response of the first floor is maximised during the excitation period. However, after the ground motion stops, a plastic deformation of almost 0.37% (17 mm) is observed.

It should be noted that the ground floor is almost a third longer than the rest of the floors and the ground floor columns have the same cross section as the rest of the building columns. Increasing the ground floor column cross sections could reduce the resulting drift to some extent. The maximum roof drift is also calculated for all eight records and shown in The envelope value is found to be 0.98%, which occurs due to ground motion record (S3).

A collection of sixteen actual records are used in the nonlinear dynamic analysis. Some of these records were selected form the seismic database of McMaster University, Canada (Naumoski *et al.*, 1988) and other records were selected from the database of Pacific Earthquake Engineering Research Centre (PEER, 2007). The records were selected based on the peak spectral acceleration to peak velocity ratio (A/V ratio) similar to the seismic motion is Vancouver, which is close to 1. These records are scaled based on the seismic hazard spectrum for Vancouver.

The characteristics of the actual records used in the analysis before scaling are shown in, where, A and V represent the spectral acceleration and velocity of a GMR, respectively. As scaling affects the dynamic response of the structure (Naumoski *et al.*, 2004), the following two methods of scaling are used here to identify the sensitivity of the response to scaling: the ordinate method and the partial area method First, spectral analysis is performed for each record. The response spectrum is then scaled to match the design spectrum of Vancouver.

The ordinate method of scaling is performed based on the fundamental period of vibration of the structure, as explained here with reference to. The response spectral acceleration corresponding to the fundamental period (Sa_2) is scaled to the value of the design spectral acceleration (Sa_1) corresponding to the same period. In other words all data points of the record are scaled based on the factor Sa_1/Sa_2.

On the other hand, the partial area method of record scaling) is based on the first and second period of vibration of the structure. The area (A_2) under the response spectral acceleration curve between 1.2 times the fundamental period value and the second period value is scaled to the corresponding area (A_1) under the design spectral acceleration curve between the same periods. All the data points of the record are scaled based on the factor A_1/A_2. The response spectra of the real GMRs scaled using the above two methods for the six-storey building is shown in and compared with the NBCC 2005 design spectrum.

Shows the maximum inter-storey drifts of the building frame produced by IDARC2D. The maximum Mean+SD values of the inter-storey drift are 1.23 and 1.26% for the ordinate method and partial area method scaled records, respectively, which occur at the first storey level. Similar values have been produced using DRAIN2DX program as shown in, which follow the same trend and orders of magnitudes for the inter-storey drift (mean+SD). The difference in the maximum dynamic drift demand values calculated using IDARC2D and DRAIN2DX are within 10% in all cases.

- The article presents a pre and a post processor for the IDARC2D computer program that is used for inelastic dynamic analysis of reinforced concrete buildings. The tools developed herein are simple and easy to use, so that the user can concentrate on the analysis rather than troubleshooting the data file construction for the analysis program.
- The pre-processor has been developed using Visual Basic operated on an Excel workbook, while the post-processor is based on the MATLAB environment.
- The pre and post processing tools developed here have been demonstrated through a six storey RC frame building which

has been designed according to the National Building Code of Canada.

- The building is assumed to be located in Vancouver representing a high level of seismic hazard. NBCC 2005 seismic provisions have been used in the design. After the design phase, the building has been analysed against eight synthetic and sixteen real ground acceleration records corresponding to the seismicity in Vancouver to determine its performance characteristics.
- The building is expected to resist collapse when subjected to this level of seismic hazard and the maximum inter-storey drift value should not exceed 2.5%. Since no inter-storey drift values exceeds 2.5% and the structure did not collapse then the design is satisfactory.
- The IDARC2D results have been verified with that produced by DRAIN2DX. For DRAIN2DX analysis, the RC elements have been modelled using simple beam-column elements and to generate the properties of these elements based on the section size and reinforcement details, a VB-Excel tool has been developed. The results produced by IDARC2D and DRAIN2DX have been found to be comparable.

The FRDP research support provided by the Concordia University to the second author is gratefully acknowledged. The research presented here forms a part of the M.A.Sc. thesis at Concordia University by the first author under the supervision of the second author. The authors will also like to thank Mr. Wael Atassi, a graduate student at Concordia University for his valuable input

The Analysis of Seismic and Temperature Structural Loads in All-Steel Buildings

There are additional considerations, rather than that of rain, snow, and wind loading, that will impact the cohesion of any all-steel structure. These involve temperature loads as well as seismic (or earthquake) loads.

The destruction generated by a mighty earthquake on existing structures can be a sobering warning of what nature can inflict on manufactured structures. Once more is understood about seismic action, the more that construction standards are adjusted to estimate resistance and defection in a building to this activity.

There are a couple of ideas of examining earthquake generation and its impact on structures. One holds that the majority of earthquakes begin when a couple of parcels of the earth's crust abut or move against one another. Ground agitation commences on the surface and initiates seismic shock waves. From the centre of the quake all of these seismic waves will decline in intensity.

Earthquake forces are carried by the inertia of a structure that is not affected by any surface movement, states another belief. The heavier the structure, the greater the seismic hock wave that impacts it. The bottom of the building goes along as the ground shifts away from the structure, yet inertia keeps the rest of the building in place for a while.

The extent to which seismic activity can jeopardize a building is caused by many factors. The type of land that the structure stands upon is critical. There is an augmentation in the amount of seismic effects on a structure with particular soils. The quantity of building firmness is also a consideration. Planned resistance to any seismic force is critical for any structure's endurance consisting of the lateral load resisting features that have been engineered into the structure.

Recent seismic resistant building engineering is centred around the premise of ductility, or the ability of the structure to have key reinforcing components deform but not break. For building code provisions having to do with seismic activity to be pertinent the main influence is ductility. The correct applications of seismic codes should help any structure in going through major earthquakes without a structural cave-in, moderate earthquakes with no major structural damage, and minimal earthquakes with no damage.

Steel will contract and enlarge as the ambient thermal conditions increases and decreases and that is why heat and cold loads are important to include in pre-engineered steel structure assembly. Mostly, temperature loads are a result of the addition of the climate, level of insulation, and building use.

Coming up with the right thermal loads for smaller steel buildings, structures in moderate climates, or facilities with climate control, may not be needed. For non-heated one level steel structures with wide free-span capacity and also where there are big differences in temperature seasonally, however, it may be required. Thermal contraction due to cold conditions, for example, may damage bolts or welds on all-steel structures. If there is at a minimum an expectancy of an increase or a decrease of 50 degrees Fahrenheit from the anticipated temperature at the point of the structure's assembly then cold and heat loading estimations should be used in steel building designs.

9

Sewage Treatment

Sewage treatment, or domestic wastewater treatment, is the process of removing contaminants from wastewater and household sewage, both runoff (effluents) and domestic. It includes physical, chemical, and biological processes to remove physical, chemical and biological contaminants. Its objective is to produce an environmentally-safe fluid waste stream (or treated effluent) and a solid waste (or treated sludge) suitable for disposal or reuse (usually as farm fertilizer). Using advanced technology it is now possible to re-use sewage effluent for drinking water, although Singapore is the only country to implement such technology on a production scale in its production of NEWater.

Origins of Sewage

Sewage is generated by residential, institutional, and commercial and industrial establishments. It includes household waste liquid from toilets, baths, showers, kitchens, sinks and so forth that is disposed of via sewers. In many areas, sewage also includes liquid waste from industry and commerce. The separation and draining of household waste into greywater and blackwater is becoming more common in the developed world, with greywater being permitted to be used for watering plants or recycled for flushing toilets.

Sewage may include stormwater runoff. Sewerage systems capable of handling stormwater are known as combined systems. Combined sewer systems are usually avoided now because precipitation causes widely varying flows reducing sewage treatment plant efficiency. Combined sewers require much larger and more expensive treatment facilities than sanitary sewers. Heavy storm runoff may overwhelm the sewage treatment system, causing a spill or overflow. Sanitary sewers are typically much smaller than combined sewers, and they are not designed to transport stormwater. Backups of raw sewage can occur if excessive Infiltration/Inflow is allowed into a sanitary sewer system.

Modern sewered developments tend to be provided with separate storm drain systems for rainwater. As rainfall travels over roofs and the ground, it may pick up various contaminants including soil particles and other sediment, heavy metals, organic compounds, animal waste, and oil and grease. Some jurisdictions require stormwater to receive some level of treatment before being discharged directly into waterways. Examples of treatment processes used for stormwater include retention basins, wetlands, buried vaults with various kinds of media filters, and vortex separators (to remove coarse solids).

Process Overview

Sewage can be treated close to where it is created, a decentralised system, (in septic tanks, biofilters or aerobic treatment systems), or be collected and transported via a network of pipes and pump stations to a municipal treatment plant, a centralised system. Sewage collection and treatment is typically subject to local, state and federal regulations and standards. Industrial sources of wastewater often require specialized treatment processes.

Sewage treatment generally involves three stages, called primary, secondary and tertiary treatment.

- *Primary treatment* consists of temporarily holding the sewage in a quiescent basin where heavy solids can settle to the bottom while oil, grease and lighter solids float to the surface. The settled and floating materials are removed and the remaining liquid may be discharged or subjected to secondary treatment.
- *Secondary treatment* removes dissolved and suspended biological matter. Secondary treatment is typically performed by indigenous, water-borne micro-organisms in a managed habitat. Secondary treatment may require a separation process to remove the micro-organisms from the treated water prior to discharge or tertiary treatment.
- *Tertiary treatment* is sometimes defined as anything more than primary and secondary treatment in order to allow rejection into a highly sensitive or fragile ecosystem (estuaries, low-flow rivers, coral reefs,...). Treated water is sometimes

disinfected chemically or physically (for example, by lagoons and microfiltration) prior to discharge into a stream, river, bay, lagoon or wetland, or it can be used for the irrigation of a golf course, green way or park. If it is sufficiently clean, it can also be used for groundwater recharge or agricultural purposes.

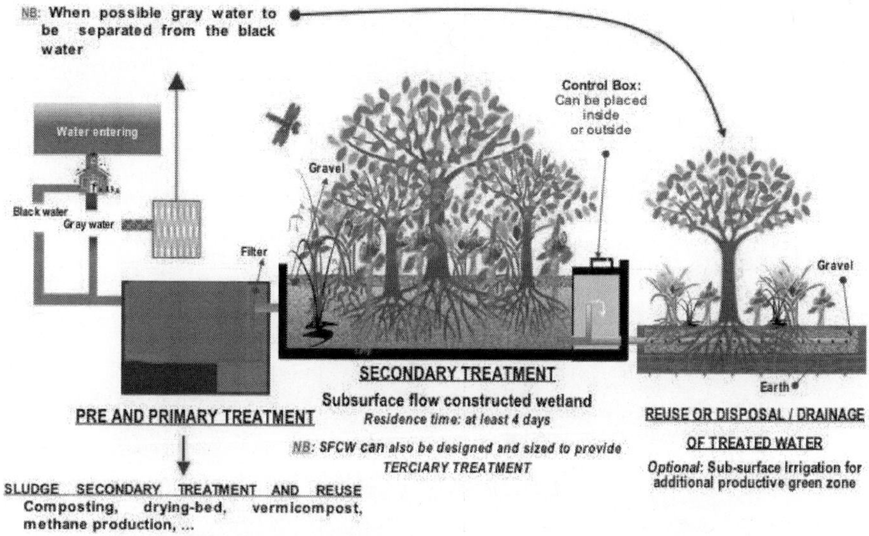

Figure: *Process Flow Diagram for a typical treatment plant via Subsurface Flow Constructed Wetlands (SFCW)*

Pre-treatment

Pre-treatment removes materials that can be easily collected from the raw waste water before they damage or clog the pumps and skimmers of primary treatment clarifiers (trash, tree limbs, leaves, etc.).

Screening

The influent sewage water is screened to remove all large objects like cans, rags, sticks, plastic packets etc. carried in the sewage stream. This is most commonly done with an automated mechanically raked bar screen in modern plants serving large populations, whilst in smaller or less modern plants a manually cleaned screen may be used. The raking action of a mechanical bar screen is typically paced according to the accumulation on the bar screens and/or flow rate. The solids are collected and later disposed in a landfill or incinerated. Bar screens or mesh screens of varying sizes may be used to optimize solids removal. If gross solids are not removed they become entrained in pipes and moving parts of the treatment plant and can cause substantial damage and inefficiency in the process.

Grit Removal

Pre-treatment may include a sand or grit channel or chamber where the velocity of the incoming wastewater is adjusted to allow the settlement of sand, grit, stones, and broken glass. These particles are removed because they may damage pumps and other equipment. For small sanitary sewer systems, the grit chambers may not be necessary, but grit removal is desirable at larger plants.

Figure: *An empty sedimentation tank at the treatment plant in Merchtem, Belgium.*

Flow Equalization

Clarifiers and mechanized secondary treatment are more efficient under uniform flow conditions. Equalization basins may be used for temporary storage of diurnal or wet-weather flow peaks. Basins provide a place to temporarily hold incoming sewage during plant maintenance and a means of diluting and distributing batch discharges of toxic or high-strength waste which might otherwise inhibit biological secondary treatment (including portable toilet waste, vehicle holding tanks, and septic tank pumpers). Flow equalization basins require variable discharge control, typically include provisions for bypass and cleaning, and may also include aerators. Cleaning may be easier if the basin is downstream of screening and grit removal.

Fat and Grease Removal

In some larger plants, fat and grease is removed by passing the sewage through a small tank where skimmers collect the fat floating on the surface. Air blowers in the base of the tank may also be used to help recover the fat as a froth. Many plants, however, use primary clarifiers with mechanical surface skimmers for fat and grease removal.

Primary Treatment

In the primary sedimentation stage, sewage flows through large tanks, commonly called "primary clarifiers" or "primary sedimentation tanks." The tanks are used to settle sludge while grease and oils rise to the surface and are skimmed off. Primary settling tanks are usually equipped with mechanically driven scrapers that continually drive the collected sludge towards a hopper in the base of the tank where it is pumped to sludge treatment facilities. Grease and oil from the floating material can sometimes be recovered for saponification.

The dimensions of the tank should be designed to effect removal of a high percentage of the floatables and sludge. A typical sedimentation tank may remove from 50 to 70 percent of suspended solids, and from 30 to 35 percent of biochemical oxygen demand (BOD) from the sewage.

Secondary Treatment

Secondary treatment is designed to substantially degrade the biological content of the sewage which are derived from human waste, food waste, soaps and detergent. The majority of municipal plants treat the settled sewage liquor using aerobic biological processes. To be effective, the biota require both oxygen and food to live. The bacteria and protozoa consume biodegradable soluble organic contaminants (e.g. sugars, fats, organic short-chain carbon molecules, etc.) and bind much of the less soluble fractions into floc. Secondary treatment systems are classified as *fixed-film* or *suspended-growth* systems.

- Fixed-film or attached growth systems include trickling filters, biotowers, and rotating biological contactors, where the biomass grows on media and the sewage passes over its surface. The fixed-film principal has further developed into Moving Bed Biofilm Reactors (MBBR), and Integrated Fixed-Film Activated Sludge (IFAS) processes. An MBBR system typically requires smaller footprint than suspended-growth systems. (Black & Veatch)

- Suspended-growth systems include activated sludge, where the biomass is mixed with the sewage and can be operated in a smaller space than trickling filters that treat the same amount of water. However, fixed-film systems are more able to cope with drastic changes in the amount of biological material and can provide higher removal rates for organic material and suspended solids than suspended growth systems.

Roughing filters are intended to treat particularly strong or variable organic loads, typically industrial, to allow them to then be treated by conventional secondary treatment processes. Characteristics include filters filled with media to which wastewater is applied. They are designed to allow high hydraulic loading and a high level of aeration. On larger installations, air is forced through the media using blowers. The resultant wastewater is usually within the normal range for conventional treatment processes.

A filter removes a small percentage of the suspended organic matter, while the majority of the organic matter undergoes a change of character, only due to the biological oxidation and nitrification taking place in the filter. With this aerobic oxidation and nitrification, the organic solids are converted into coagulated suspended mass, which is heavier and bulkier, and can settle to the bottom of a tank. The effluent of the filter is therefore passed through a sedimentation tank, called a secondary clarifier, secondary settling tank or humus tank.

Activated Sludge

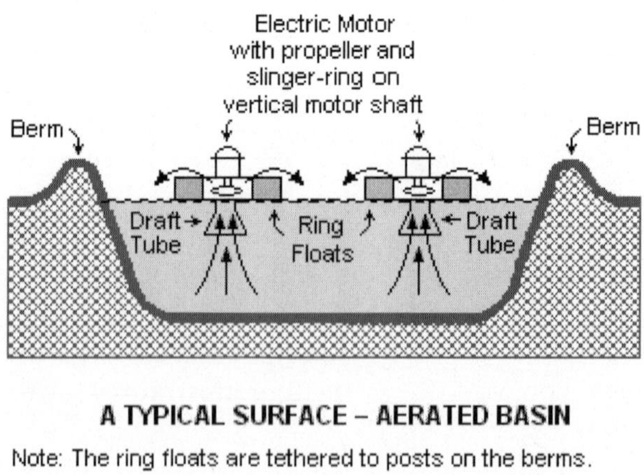

Figure: *A Typical Surface-Aerated Basin (using motor-driven floating aerators)*

In general, activated sludge plants encompass a variety of mechanisms and processes that use dissolved oxygen to promote the growth of biological floc that substantially removes organic material. The process traps particulate material and can, under ideal conditions, convert ammonia to nitrite and nitrate ultimately to nitrogen gas.

Surface-aerated Basins (Lagoons)

Many small municipal sewage systems in the United States (1 million gal./day or less) use aerated lagoons. Most biological oxidation processes for treating industrial wastewaters have in common the use of oxygen (or air) and microbial action. Surface-aerated basins achieve 80 to 90 percent removal of BOD with retention times of 1 to 10 days. The basins may range in depth from 1.5 to 5.0 metres and use motor-driven aerators floating on the surface of the wastewater.

In an aerated basin system, the aerators provide two functions: they transfer air into the basins required by the biological oxidation reactions, and they provide the mixing required for dispersing the air and for contacting the reactants (that is, oxygen, wastewater and microbes). Typically, the floating surface aerators are rated to deliver the amount of air equivalent to 1.8 to 2.7 kg O_2/kW·h. However, they do not provide as good mixing as is normally achieved in activated sludge systems and therefore aerated basins do not achieve the same performance level as activated sludge units.

Biological oxidation processes are sensitive to temperature and, between 0 °C and 40 °C, the rate of biological reactions increase with temperature. Most surface aerated vessels operate at between 4 °C and 32 °C.

Constructed Wetlands

Constructed wetlands (can either be surface flow or subsurface flow, horizontal or vertical flow), include engineered reedbeds and belong to the family of phytorestoration and ecotechnologies; they provide a high degree of biological improvement and depending on design, act as a primary, secondary and sometimes tertiary treatment. One example is a small reedbed used to clean the drainage from the elephants' enclosure at Chester Zoo in England; numerous CWs are used to recycle the water of the city of Honfleur in France and numerous other towns in Europe, the US, Asia and Australia. They are known to be highly productive systems as they copy natural wetlands, called the "Kidneys of the earth" for their fundamental recycling capacity of the hydrological cycle in the biosphere. Robust and reliable, their treatment capacities improve as time go by, at the opposite of conventional treatment plants whose machinery age with time. They are being increasingly used, although adequate and experienced design are more fundamental than for other systems and space limitation may impede their use.

Filter Beds (Oxidizing Beds)

In older plants and those receiving variable loadings, trickling filter beds are used where the settled sewage liquor is spread onto the surface of a bed made up of coke (carbonized coal), limestone chips or specially fabricated plastic media. Such media must have large surface areas to support the biofilms that form. The liquor is typically distributed through perforated spray arms. The distributed liquor trickles through the bed and is collected in drains at the base. These drains also provide a source of air which percolates up through the bed, keeping it aerobic. Biological films of bacteria, protozoa and fungi form on the media's surfaces and eat or otherwise reduce the organic content. This biofilm is often grazed by insect larvae, snails, and worms which help maintain an optimal thickness. Overloading of beds increases the thickness of the film leading to clogging of the filter media and ponding on the surface. Recent advances in media and process micro-biology design overcome many issues with Trickling filter designs.

Soil Bio-technology

A new process called soil bio-technology (SBT) developed at IIT Bombay has shown tremendous improvements in process efficiency enabling total water reuse, due to extremely low operating power requirements of less than 50 joules per kg of treated water. Typically SBT systems can achieve chemical oxygen demand (COD) levels less than 10 mg/L from sewage input of COD 400 mg/L. SBT plants exhibit high reductions in COD values and bacterial counts as a result of the very high microbial densities available in the media. Unlike conventional treatment plants, SBT plants produce insignificant amounts of sludge, precluding the need for sludge disposal areas that are required by other technologies.

In the Indian context, conventional sewage treatment plants fall into systemic disrepair due to 1) high operating costs, 2) equipment corrosion due to methanogenesis and hydrogen sulphide, 3) non-reusability of treated water due to high COD (>30 mg/L) and high fecal coliform (>3000 NFU) counts, 4) lack of skilled operating personnel and 5) equipment replacement issues. Examples of such systemic failures has been documented by Sạnkat Mochan Foundation at the Ganges basin after a massive cleanup effort by the Indian government in 1986 by setting up sewage treatment plants under the Ganga Action Plan failed to improve river water quality.

Biological Aerated Filters

Biological Aerated (or Anoxic) Filter (BAF) or Biofilters combine filtration with biological carbon reduction, nitrification or denitrification. BAF usually includes a reactor filled with a filter media. The media is either in suspension or supported by a gravel layer at the foot of the filter. The dual purpose of this media is to support highly active biomass that is attached to it and to filter suspended solids. Carbon reduction and ammonia conversion occurs in aerobic mode and sometime achieved in a single reactor while nitrate conversion occurs in anoxic mode. BAF is operated either in upflow or downflow configuration depending on design specified by manufacturer.

Rotating Biological Contactors

Rotating biological contactors (RBCs) are mechanical secondary treatment systems, which are robust and capable of withstanding surges in organic load. RBCs were first installed in Germany in 1960 and have since been developed and refined into a reliable operating unit. The rotating disks support the growth of bacteria and micro-organisms present in the sewage, which break down and stabilise organic pollutants. To be successful, micro-organisms need both oxygen to live and food to grow. Oxygen is obtained from the atmosphere as the disks rotate. As the micro-organisms grow, they build up on the media until they are sloughed off due to shear forces provided by the rotating discs in the sewage. Effluent from the RBC is then passed through final clarifiers where the micro-organisms in suspension settle as a sludge. The sludge is withdrawn from the clarifier for further treatment. A functionally similar biological filtering system has become popular as part of home aquarium filtration and purification. The aquarium water is drawn up out of the tank and then cascaded over a freely spinning corrugated fiber-mesh wheel before passing through a media filter and back into the aquarium. The spinning mesh wheel develops a biofilm coating of microorganisms that feed on the suspended wastes in the aquarium water and are also exposed to the atmosphere as the wheel rotates. This is especially good at removing waste urea and ammonia urinated into the aquarium water by the fish and other animals.

Membrane Bioreactors

Membrane bioreactors (MBR) combine activated sludge treatment with a membrane liquid-solid separation process. The membrane component uses low pressure microfiltration or ultrafiltration membranes and eliminates the need for clarification and tertiary filtration. The membranes are typically immersed in the aeration tank; however, some applications utilize a separate membrane tank. One of the key benefits of an MBR system is that it effectively overcomes the limitations associated with poor settling of sludge in conventional activated sludge (CAS) processes. The technology permits bioreactor operation with considerably higher mixed liquor suspended solids (MLSS) concentration than CAS systems, which are limited by sludge settling. The process is typically operated at MLSS in the range of 8,000–12,000 mg/L, while CAS are operated in the range of 2,000–3,000 mg/L. The elevated biomass concentration in the MBR process allows for very effective removal of both soluble and particulate biodegradable materials at higher loading rates. Thus increased sludge retention times, usually exceeding 15 days, ensure complete nitrification even in extremely cold weather.

The cost of building and operating an MBR is usually higher than conventional wastewater treatment. Membrane filters can be blinded with grease or abraded by suspended grit and lack a clarifier's flexibility to pass peak flows. The technology has become increasingly popular for reliably pretreated waste streams and has gained wider acceptance where infiltration and inflow have been controlled, however, and the life-cycle costs have been steadily decreasing. The small footprint of MBR systems, and the high quality effluent produced, make them particularly useful for water reuse applications.

Secondary Sedimentation

The final step in the secondary treatment stage is to settle out the biological floc or filter material through a secondary clarifier and to produce sewage water containing low levels of organic material and suspended matter.

Tertiary Treatment

The purpose of tertiary treatment is to provide a final treatment stage to raise the effluent quality before it is discharged to the receiving environment (sea, river, lake, ground, etc.). More than one tertiary treatment process may be used at any treatment plant. If disinfection is practiced, it is always the final process. It is also called "effluent polishing."

Filtration

Sand filtration removes much of the residual suspended matter. Filtration over activated carbon, also called *carbon adsorption,* removes residual toxins.

Lagooning

Lagooning provides settlement and further biological improvement through storage in large man-made ponds or lagoons. These lagoons are highly aerobic and colonization by native macrophytes, especially reeds, is often encouraged. Small filter feeding invertebrates such as *Daphnia* and species of *Rotifera* greatly assist in treatment by removing fine particulates.

Nutrient Removal

Wastewater may contain high levels of the nutrients nitrogen and phosphorus. Excessive release to the environment can lead to a build up of nutrients, called eutrophication, which can in turn encourage the overgrowth of weeds, algae, and cyanobacteria (blue-green algae). This may cause an algal bloom, a rapid growth in the population of algae. The algae numbers are unsustainable and eventually most of them die. The decomposition of the algae by bacteria uses up so much of oxygen in the water that most or all of the animals die, which creates more organic matter for the bacteria to decompose. In addition to causing deoxygenation, some algal species produce toxins that contaminate drinking water supplies. Different treatment processes are required to remove nitrogen and phosphorus.

Nitrogen Removal

The removal of nitrogen is effected through the biological oxidation of nitrogen from ammonia to nitrate (nitrification), followed by denitrification, the reduction of nitrate to nitrogen gas. Nitrogen gas is released to the atmosphere and thus removed from the water.

Nitrification itself is a two-step aerobic process, each step facilitated by a different type of bacteria. The oxidation of ammonia (NH_3) to nitrite (NO_2^-) is most often facilitated by *Nitrosomonas* spp. (nitroso referring to the formation of a nitroso functional group). Nitrite oxidation to nitrate (NO_3^-), though traditionally believed to be facilitated by *Nitrobacter* spp. (nitro referring the formation of a nitro functional group), is now known to be facilitated in the environment almost exclusively by *Nitrospira* spp. Denitrification requires anoxic conditions to encourage the appropriate biological communities to form. It is facilitated by a wide diversity of bacteria. Sand filters, lagooning and reed beds can all be used to reduce nitrogen, but the activated sludge process (if designed well) can do the job the most easily. Since denitrification is the reduction of nitrate to dinitrogen gas, an electron donor is needed. This can be, depending on the wastewater, organic matter (from faeces), sulfide, or an added donor like methanol. The sludge in the anoxic tanks (denitrification tanks) must be mixed well (mixture of recirculated mixed liquor, return activated sludge [RAS], and raw influent) e.g. by using submersible mixers in order to achieve the desired denitrification. Sometimes the conversion of toxic ammonia to nitrate alone is referred to as tertiary treatment.

Many sewage treatment plants use axial flow pumps to transfer the nitrified mixed liquor from the aeration zone to the anoxic zone for denitrification. These pumps are often referred to as *Internal Mixed Liquor Recycle* (IMLR) pumps.

Phosphorus Removal

Each person excretes between 200 and 1000 grams of phosphorus annually. Studies of United States sewage in the late 1960s estimated mean per capita contributions of 500 grams in urine and feces, 1000 grams in synthetic detergents, and lesser variable amounts used as corrosion and scale control chemicals in water supplies. Source control via alternative detergent formulations has subsequently reduced the largest contribution, but the content of urine and feces will remain unchanged. Phosphorus removal is important as it is a limiting nutrient for algae growth in many fresh water systems. (For a description of the negative effects of algae). It is also particularly important for water reuse systems where high phosphorus concentrations may lead to fouling of downstream equipment such as reverse osmosis.

Phosphorus can be removed biologically in a process called enhanced biological phosphorus removal. In this process, specific bacteria, called polyphosphate accumulating organisms (PAOs), are selectively enriched and accumulate large quantities of phosphorus within their cells (up to 20 percent of their mass). When the biomass enriched in these bacteria is separated from the treated water, these biosolids have a high fertilizer value.

Phosphorus removal can also be achieved by chemical precipitation, usually with salts of iron (e.g. ferric chloride), aluminum (e.g. alum), or lime. This may lead to excessive sludge production as hydroxides precipitates and the added chemicals can be expensive. Chemical phosphorus removal requires significantly smaller equipment footprint than biological removal, is easier to operate and is often more reliable than biological phosphorus removal. Another method for phosphorus removal is to use granular laterite.

Once removed, phosphorus, in the form of a phosphate-rich sludge, may be stored in a land fill or resold for use in fertilizer.

Disinfection

The purpose of disinfection in the treatment of waste water is to substantially reduce the number of microorganisms in the water to be discharged back into the environment for the later use of drinking, bathing, irrigation, etc. The effectiveness of disinfection depends on the quality of the water being treated (e.g., cloudiness, pH, etc.), the type of disinfection being used, the disinfectant dosage (concentration and time), and other environmental variables. Cloudy water will be treated less successfully, since solid matter can shield organisms, especially from ultraviolet light or if contact times are low. Generally, short contact times, low doses and high flows all militate against effective disinfection. Common methods of disinfection include ozone, chlorine, ultraviolet light, or sodium hypochlorite. Chloramine, which is used for drinking water, is not used in waste water treatment because of its persistence. After multiple steps of disinfection, the treated water is ready to be released back into the water cycle by means of the nearest body of water or agriculture. Afterwards, the water can be transferred to reserves for everyday human uses.

Chlorination remains the most common form of waste water disinfection in North America due to its low cost and long-term history of effectiveness. One disadvantage is that chlorination of residual organic material can generate chlorinated-organic compounds that may be carcinogenic or harmful to the environment. Residual chlorine or chloramines may also be capable of chlorinating organic material in the natural aquatic environment. Further, because residual chlorine is toxic to aquatic species, the treated effluent must also be chemically dechlorinated, adding to the complexity and cost of treatment.

Ultraviolet (UV) light can be used instead of chlorine, iodine, or other chemicals. Because no chemicals are used, the treated water has no adverse effect on organisms that later consume it, as may be the case with other methods. UV radiation causes damage to the genetic structure of bacteria, viruses, and other pathogens, making them incapable of reproduction. The key disadvantages of UV disinfection are the need for frequent lamp maintenance and replacement and the need for a highly treated effluent to ensure that the target microorganisms are not shielded from the UV radiation (i.e., any solids present in the treated effluent may protect microorganisms from the UV light). In the United Kingdom, UV light is becoming the most common means of disinfection because of the concerns about the impacts of chlorine in chlorinating residual organics in the wastewater and in chlorinating organics in the receiving water. Some sewage treatment systems in Canada and the US also use UV light for their effluent water disinfection.

Ozone (O_3) is generated by passing oxygen (O_2) through a high voltage potential resulting in a third oxygen atom becoming attached and forming O_3. Ozone is very unstable and reactive and oxidizes most organic material it comes in contact with, thereby destroying many pathogenic microorganisms. Ozone is considered to be safer than chlorine because, unlike chlorine which has to be stored on site (highly poisonous in the event of an accidental release), ozone is generated onsite as needed. Ozonation also produces fewer disinfection by-products than chlorination. A disadvantage of ozone disinfection is the high cost of the ozone generation equipment and the requirements for special operators.

Odor Control

Odors emitted by sewage treatment are typically an indication of an anaerobic or "septic" condition. Early stages of processing will tend to produce foul smelling gases, with hydrogen sulfide being most common in generating complaints. Large process plants in urban areas will often treat the odors with carbon reactors, a contact media with bio-slimes, small doses of chlorine, or circulating fluids to biologically capture and metabolize the obnoxious gases. Other methods of odor control exist, including addition of iron salts, hydrogen peroxide, calcium nitrate, etc. to manage hydrogen sulfide levels. High-density solids pumps are suitable to reduce odors by conveying sludge through hermetic closed pipework.

Package Plants and Batch Reactors

To use less space, treat difficult waste and intermittent flows, a number of designs of hybrid treatment plants have been produced. Such plants often combine at least two stages of the three main treatment stages into one combined stage. In the UK, where a large number of wastewater treatment plants serve small populations, package plants are a viable alternative to building a large structure for each process stage. In the US, package plants are typically used in rural areas, highway rest stops and trailer parks.

One type of system that combines secondary treatment and settlement is the sequencing batch reactor (SBR). Typically, activated sludge is mixed with raw incoming sewage, and then mixed and aerated. The settled sludge is run off and re-aerated before a proportion is returned to the headworks. SBR plants are now being deployed in many parts of the world.

The disadvantage of the SBR process is that it requires a precise control of timing, mixing and aeration. This precision is typically achieved with computer controls linked to sensors. Such a complex, fragile system is unsuited to places where controls may be unreliable, poorly maintained, or where the power supply may be intermittent. Extended aeration package plants use separate basins for aeration and settling, and are somewhat larger than SBR plants with reduced timing sensitivity.

Package plants may be referred to as *high charged* or *low charged*. This refers to the way the biological load is processed. In high charged systems, the biological stage is presented with a high organic load and the combined floc and organic material is then oxygenated for a few hours before being charged again with a new load. In the low charged system the biological stage contains a low organic load and is combined with flocculate for longer times.

Sludge Treatment and Disposal

The sludges accumulated in a wastewater treatment process must be treated and disposed of in a safe and effective manner. The purpose of digestion is to reduce the amount of organic matter and the number of disease-causing microorganisms present in the solids. The most common treatment options include anaerobic digestion, aerobic digestion, and composting. Incineration is also used albeit to a much lesser degree. Sludge treatment depends on the amount of solids generated and other site-specific conditions. Composting is most often applied to small-scale plants with aerobic digestion for mid sized operations, and anaerobic digestion for the larger-scale operations.

Anaerobic Digestion

Anaerobic digestion is a bacterial process that is carried out in the absence of oxygen. The process can either be *thermophilic* digestion, in which sludge is fermented in tanks at a temperature of 55°C, or *mesophilic*, at a temperature of around 36°C. Though allowing shorter retention time (and thus smaller tanks), thermophilic digestion is more expensive in terms of energy consumption for heating the sludge.

Anaerobic digestion is the most common (mesophilic) treatment of domestic sewage in septic tanks, which normally retain the sewage from one day to two days, reducing the BOD by about 35 to 40 percent. This reduction can be increased with a combination of anaerobic and aerobic treatment by installing *Aerobic Treatment Units* (ATUs) in the septic tank.

One major feature of anaerobic digestion is the production of biogas (with the most useful component being methane), which can be used in generators for electricity production and/or in boilers for heating purposes.

Aerobic Digestion

Aerobic digestion is a bacterial process occurring in the presence of oxygen. Under aerobic conditions, bacteria rapidly consume organic matter and convert it into carbon dioxide. The operating costs used to be characteristically much greater for aerobic digestion because of the energy used by the blowers, pumps and motors needed to add oxygen to the process. Aerobic digestion can also be achieved by using diffuser systems or jet aerators to oxidize the sludge. Fine bubble diffusers are typically the more cost-efficient diffusion method, however, plugging is typically a problem due to sediment settling into the smaller air holes. Coarse bubble diffusers are more commonly used in activated sludge tanks (generally a side process in waste water management) or in the flocculation stages. A key component for selecting diffuser type is to ensure it will produce the required oxygen transfer rate.

Composting

Composting is also an aerobic process that involves mixing the sludge with sources of carbon such as sawdust, straw or wood chips. In the presence of oxygen, bacteria digest both the wastewater solids and the added carbon source and, in doing so, produce a large amount of heat.

Incineration

Incineration of sludge is less common because of air emissions concerns and the supplemental fuel (typically natural gases or fuel oil) required to burn the low calorific value sludge and vaporize residual water. Stepped multiple hearth incinerators with high residence time and fluidized bed incinerators are the most common systems used to combust wastewater sludge. Co-firing in municipal waste-to-energy plants is occasionally done, this option being less expensive assuming the facilities already exist for solid waste and there is no need for auxiliary fuel.

Sludge Disposal

When a liquid sludge is produced, further treatment may be required to make it suitable for final disposal. Typically, sludges are thickened (dewatered) to reduce the volumes transported off-site for disposal. There is no process which completely eliminates the need to dispose of biosolids. There is, however, an additional step some cities are taking to superheat sludge and convert it into small pelletized granules that are high in nitrogen and other organic materials. In New York City, for example, several sewage treatment plants have dewatering facilities that use large centrifuges along with the addition of chemicals such as polymer to further remove liquid from the sludge. The removed fluid, called centrate, is typically reintroduced into the wastewater process. The product which is left is called "cake" and that is picked up by companies which turn it into fertilizer pellets. This product is then sold to local farmers and turf farms as a soil amendment or fertilizer, reducing the amount of space required to dispose of sludge in landfills. Much sludge originating from commercial or industrial areas is contaminated with toxic materials that are released into the sewers from the industrial processes. Elevated concentrations of such materials may make the sludge unsuitable for agricultural use and it may then have to be incinerated or disposed of to landfill.

Treatment in the Receiving Environment

Figure: *The outlet of the Karlsruhe sewage treatment plant flows into the Alb.*

Many processes in a wastewater treatment plant are designed to mimic the natural treatment processes that occur in the environment, whether that environment is a natural water body or the ground. If not overloaded, bacteria in the environment will consume organic contaminants, although this will reduce the levels of oxygen in the water and may significantly change the overall ecology of the receiving water. Native bacterial populations feed on the organic contaminants, and the numbers of disease-causing microorganisms are reduced by natural environmental conditions such as predation or exposure to ultraviolet radiation. Consequently, in cases where the receiving environment provides a high level of dilution, a high degree of wastewater treatment may not be required. However, recent evidence has demonstrated that very low levels of specific contaminants in wastewater, including hormones (from animal husbandry and residue from human hormonal contraception methods) and synthetic materials such as phthalates that mimic hormones in their action, can have an unpredictable adverse impact on the natural biota and potentially on humans if the water is re-used for drinking water. In the US and EU, uncontrolled discharges of wastewater to the environment are not permitted under law, and strict water quality requirements are to be met, as clean drinking water is essential. A significant threat in the coming decades will be the increasing uncontrolled discharges of wastewater within rapidly developing countries.

Effects on Biology

Sewage treatment plants can have multiple effects on nutrient levels in the water that the treated sewage flows into. These effects on nutrients can have large effects on the biological life in the water in contact with the effluent. Stabilization ponds (or treatment ponds) can include any of the following:

- Oxidation ponds, which are aerobic bodies of water usually 1–2 meters in depth that receive effluent from sedimentation tanks or other forms of primary treatment.
- Dominated by algae
- Polishing ponds are similar to oxidation ponds but receive effluent from an oxidation pond or from a plant with an extended mechanical treatment.
- Dominated by zooplankton
- Facultative lagoons, raw sewage lagoons, or sewage lagoons are ponds where sewage is added with no primary treatment other than coarse screening. These ponds provide effective treatment when the surface remains aerobic; although anaerobic conditions may develop near the layer of settled sludge on the bottom of the pond.

- Anaerobic lagoons are heavily loaded ponds.
- Dominated by bacteria
- Sludge lagoons are aerobic ponds, usually 2–5 meters in depth, that receive anaerobically digested primary sludge, or activated secondary sludge under water.
- Upper layers are dominated by algae

Phosphorus limitation is a possible result from sewage treatment and results in flagellate-dominated plankton, particularly in summer and fall. At the same time a different study found high nutrient concentrations linked to sewage effluents. High nutrient concentration leads to high chlorophyll a concentrations, which is a proxy for primary production in marine environments. High primary production means high phytoplankton populations and most likely high zooplankton populations because zooplankton feed on phytoplankton. However, effluent released into marine systems also leads to greater population instability.

A study done in Britain found that the quality of effluent affected the planktonic life in the water in direct contact with the wastewater effluent. Turbid, low-quality effluents either did not contain ciliated protozoa or contained only a few species in small numbers. On the other hand, high-quality effluents contained a wide variety of ciliated protozoa in large numbers. Due to these findings, it seems unlikely that any particular component of the industrial effluent has, by itself, any harmful effects on the protozoan populations of activated sludge plants. The planktonic trends of high populations close to input of treated sewage is contrasted by the bacterial trend. In a study of *Aeromonas* spp. in increasing distance from a wastewater source, greater change in seasonal cycles was found the furthest from the effluent. This trend is so strong that the furthest location studied actually had an inversion of the *Aeromonas* spp. cycle in comparison to that of fecal coliforms. Since there is a main pattern in the cycles that occurred simultaneously at all stations it indicates seasonal factors (temperature, solar radiation, phytoplankton) control of the bacterial population. The effluent dominant species changes from *Aeromonas caviae* in winter to *Aeromonas sobria* in the spring and fall while the inflow dominant species is *Aeromonas caviae*, which is constant throughout the seasons.

Sewage Treatment in Developing Countries

Few reliable figures on the share of the wastewater collected in sewers that is being treated in the world exist. In many developing countries the bulk of domestic and industrial wastewater is discharged without any treatment or after primary treatment only. In Latin America about 15% of collected wastewater passes through treatment plants (with varying levels of actual treatment). In Venezuela, a below average country in South America with respect to wastewater treatment, 97 percent of the country's sewage is discharged raw into the environment. In a relatively developed Middle Eastern country such as Iran, the majority of Tehran's population has totally untreated sewage injected to the city's groundwater. However now the construction of major parts of the sewage system, collection and treatment, in Tehran is almost complete, and under development, due to be fully completed by the end of 2012. In Isfahan, Iran's third largest city, sewage treatment was started more than 100 years ago.

In Israel, about 50 percent of agricultural water usage (total use was 1 billion cubic metres in 2008) is provided through reclaimed sewer water. Future plans call for increased use of treated sewer water as well as more desalination plants.

Most of sub-Saharan Africa is without wastewater treatment.

Industrial Wastewater

Some industrial facilities generate ordinary domestic sewage that can be treated by municipal facilities. Industries that generate wastewater with high concentrations of conventional pollutants (e.g. oil and grease), toxic pollutants (e.g. heavy metals, volatile organic compounds) or other nonconventional pollutants such as ammonia, need specialized treatment systems. Some of these facilities can install a pre-treatment system to remove the toxic components, and then send the partially-treated wastewater to the municipal system. Industries generating large volumes of wastewater typically operate their own complete on-site treatment systems.

Some industries have been successful at redesigning their manufacturing processes to reduce or eliminate pollutants, through a process called pollution prevention.

Heated water generated by power plants or manufacturing plants may be controlled with:

- cooling ponds, man-made bodies of water designed for cooling by evaporation, convection, and radiation
- cooling towers, which transfer waste heat to the atmosphere through evaporation and/or heat transfer

- cogeneration, a process where waste heat is recycled for domestic and/or industrial heating purposes.

Industrial wastewater treatment covers the mechanisms and processes used to treat waters that have been contaminated in some way by anthropogenic industrial or commercial activities prior to its release into the environment or its re-use.

Most industries produce some wet waste although recent trends in the developed world have been to minimise such production or recycle such waste within the production process. However, many industries remain dependent on processes that produce wastewaters.

Sources of Industrial Wastewater

Agricultural Waste

Iron and Steel Industry: The production of iron from its ores involves powerful reduction reactions in blast furnaces. Cooling waters are inevitably contaminated with products especially ammonia and cyanide. Production of coke from coal in coking plants also requires water cooling and the use of water in by-products separation. Contamination of waste streams includes gasification products such as benzene, naphthalene, anthracene, cyanide, ammonia, phenols, cresols together with a range of more complex organic compounds known collectively as polycyclic aromatic hydrocarbons (PAH).

The conversion of iron or steel into sheet, wire or rods requires hot and cold mechanical transformation stages frequently employing water as a lubricant and coolant. Contaminants include hydraulic oils, tallow and particulate solids. Final treatment of iron and steel products before onward sale into manufacturing includes *pickling* in strong mineral acid to remove rust and prepare the surface for tin or chromium plating or for other surface treatments such as galvanisation or painting. The two acids commonly used are hydrochloric acid and sulfuric acid. Wastewaters include acidic rinse waters together with waste acid. Although many plants operate acid recovery plants, (particularly those using Hydrochloric acid), where the mineral acid is boiled away from the iron salts, there remains a large volume of highly acid ferrous sulfate or ferrous chloride to be disposed of. Many steel industry wastewaters are contaminated by hydraulic oil also known as *soluble oil.*

Mines and Quarries

The principal waste-waters associated with mines and quarries are slurries of rock particles in water. These arise from rainfall washing exposed surfaces and haul roads and also from rock washing and grading processes. Volumes of water can be very high, especially rainfall related arisings on large sites. Some specialized separation operations, such as coal washing to separate coal from native rock using density gradients, can produce wastewater contaminated by fine particulate haematite and surfactants. Oils and hydraulic oils are also common contaminants. Wastewater from metal mines and ore recovery plants are inevitably contaminated by the minerals present in the native rock formations. Following crushing and extraction of the desirable materials, undesirable materials may become contaminated in the wastewater. For metal mines, this can include unwanted metals such as zinc and other materials such as arsenic. Extraction of high value metals such as gold and silver may generate slimes containing very fine particles in where physical removal of contaminants becomes particularly difficult.

Food Industry

Wastewater generated from agricultural and food operations has distinctive characteristics that set it apart from common municipal wastewater managed by public or private wastewater treatment plants throughout the world: it is biodegradable and nontoxic, but that has high concentrations of biochemical oxygen demand (BOD) and suspended solids (SS). The constituents of food and agriculture wastewater are often complex to predict due to the differences in BOD and pH in effluents from vegetable, fruit, and meat products and due to the seasonal nature of food processing and postharvesting.

Processing of food from raw materials requires large volumes of high grade water. Vegetable washing generates waters with high loads of particulate matter and some dissolved organics. It may also contain surfactants.

Animal slaughter and processing produces very strong organic waste from body fluids, such as blood, and gut contents. This wastewater is frequently contaminated by significant levels of antibiotics and growth hormones from the animals and by a variety of pesticides used to control external parasites. Insecticide residues in fleeces is a particular problem in treating waters generated in wool processing.

Processing food for sale produces wastes generated from cooking which are often rich in plant organic material and may also contain salt, flavourings, colouring material and acids or alkali. Very significant quantities of oil or fats may also be present.

Complex Organic Chemicals Industry

A range of industries manufacture or use complex organic chemicals. These include pesticides, pharmaceuticals, paints and dyes, petro-chemicals, detergents, plastics, paper pollution, etc. Waste waters can be contaminated by feed-stock materials, by-products, product material in soluble or particulate form, washing and cleaning agents, solvents and added value products such as plasticisers. Treatment facilities that do not need control of their effluent typically opt for a type of aerobic treatment, i.e. Aerated Lagoons.

Nuclear Industry

The waste production from the nuclear and radio-chemicals industry is dealt with as *Radioactive waste.*

Water Treatment

Water treatment for the production of drinking water is dealt with elsewhere. Many industries have a need to treat water to obtain very high quality water for demanding purposes. Water treatment produces organic and mineral sludges from filtration and sedimentation. Ion exchange using natural or synthetic resins removes calcium, magnesium and carbonate ions from water, replacing them with hydrogen and hydroxyl ions. Regeneration of ion exchange columns with strong acids and alkalis produces a wastewater rich in hardness ions which are readily precipitated out, especially when in admixture with other wastewaters.

Treatment of Industrial Wastewater

The different types of contamination of wastewater require a variety of strategies to remove the contamination.

Solids Removal

Most solids can be removed using simple sedimentation techniques with the solids recovered as slurry or sludge. Very fine solids and solids with densities close to the density of water pose special problems. In such case filtration or ultrafiltration may be required. Although, flocculation may be used, using alum salts or the addition of polyelectrolytes.

Oils and Grease Removal

Many oils can be recovered from open water surfaces by skimming devices. Considered a dependable and cheap way to remove oil, grease and other hydrocarbons from water, oil skimmers can sometimes achieve the desired level of water purity. At other times, skimming is also a cost-efficient method to remove most of the oil before using membrane filters and chemical processes. Skimmers will prevent filters from blinding prematurely and keep chemical costs down because there is less oil to process.

Because grease skimming involves higher viscosity hydrocarbons, skimmers must be equipped with heaters powerful enough to keep grease fluid for discharge. If floating grease forms into solid clumps or mats, a spray bar, aerator or mechanical apparatus can be used to facilitate removal.

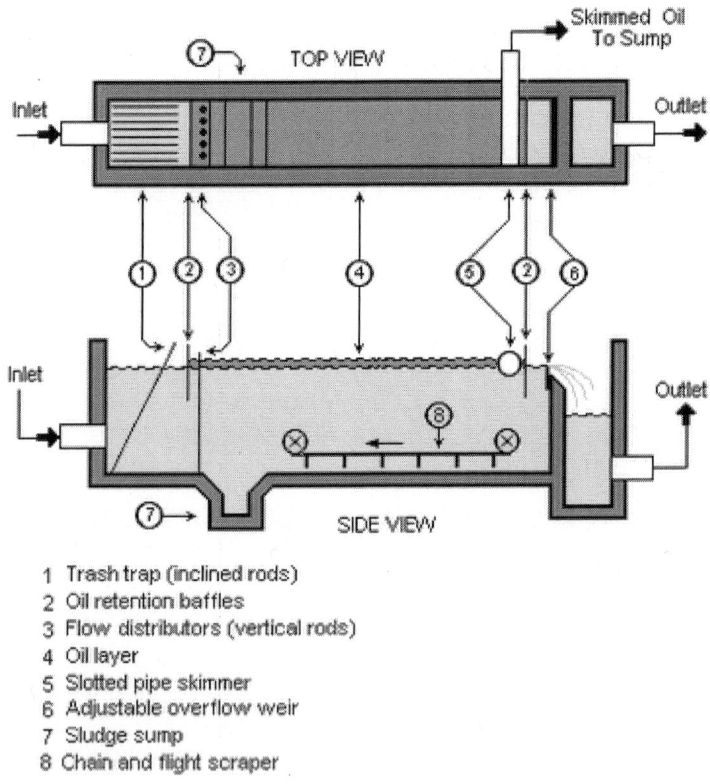

Figure: *A typical API oil-water separator used in many industries*

However, hydraulic oils and the majority of oils that have degraded to any extent will also have a soluble or emulsified component that will require further treatment to eliminate. Dissolving or emulsifying oil using surfactants or solvents usually exacerbates the problem rather than solving it, producing wastewater that is more difficult to treat. The wastewaters from large-scale industries such as oil refineries, petrochemical plants, chemical plants, and natural gas processing plants commonly contain gross amounts of oil and suspended solids. Those industries use a device known as an API oil-water separator which is designed to separate the oil and suspended solids from their wastewater effluents. The name is derived from the fact that such separators are designed according to standards published by the American Petroleum Institute (API).

The API separator is a gravity separation device designed by using Stokes Law to define the rise velocity of oil droplets based on their density and size. The design is based on the specific gravity difference between the oil and the wastewater because that difference is much smaller than the specific gravity difference between the suspended solids and water. The suspended solids settles to the bottom of the separator as a sediment layer, the oil rises to top of the separator and the cleansed wastewater is the middle layer between the oil layer and the solids.

Typically, the oil layer is skimmed off and subsequently re-processed or disposed of, and the bottom sediment layer is removed by a chain and flight scraper (or similar device) and a sludge pump. The water layer is sent to further treatment consisting usually of a Electroflotation module for additional removal of any residual oil and then to some type of biological treatment unit for removal of undesirable dissolved chemical compounds.

Parallel plate separators are similar to API separators but they include tilted parallel plate assemblies (also known as parallel packs). The parallel plates provide more surface for suspended oil droplets to coalesce into larger globules. Such separators still depend upon the specific gravity between the suspended oil and the water. However, the parallel plates enhance the degree of oil-water separation. The result is that a parallel plate separator requires significantly less space than a conventional API separator to achieve the same degree of separation.

Removal of Biodegradable Organics

Biodegradable organic material of plant or animal origin is usually possible to treat using extended conventional wastewater treatment processes such as activated sludge or trickling filter. Problems can arise if the wastewater is excessively diluted with washing water or is highly concentrated such as neat blood or milk. The presence of cleaning agents, disinfectants, pesticides, or antibiotics can have detrimental impacts on treatment processes.

Activated Sludge Process

Activated sludge is a biochemical process for treating sewage and industrial wastewater that uses air (or oxygen) and microorganisms to biologically oxidize organic pollutants, producing a waste sludge (or floc) containing the oxidized material. In general, an activated sludge process includes:

- An aeration tank where air (or oxygen) is injected and thoroughly mixed into the wastewater.
- A settling tank (usually referred to as a "clarifier" or "settler") to allow the waste sludge to settle. Part of the waste sludge is recycled to the aeration tank and the remaining waste sludge is removed for further treatment and ultimate disposal.
- As a general process for most of the Industrial waste water the following Technologies are used.

ASP: Activated Sludge process 2. SAFF system of Submerged aerobic fixed film system 3. MBBR : Moving bed bio reactor (Anox invented this now is considered generic technology) 4. MBR : Membrane Bioreactor 5. DAF clarifiers 6. TBR : Turbo bioreactor Technology (A patented technology of Wockoliver) 7. Filtration technologies More information about above can be found on various commercial manufactures like WOIL.

Trickling Filter Process

A trickling filter consists of a bed of rocks, gravel, slag, peat moss, or plastic media over which wastewater flows downward and contacts a layer (or film) of microbial slime covering the bed media. Aerobic conditions are maintained by forced air flowing through the bed or by natural convection of air. The process involves adsorption of organic compounds in the wastewater by the microbial slime layer, diffusion of air into the slime layer to provide the oxygen required for the biochemical oxidation of the organic compounds. The end products include carbon dioxide gas, water and other products of the oxidation. As the slime layer thickens, it becomes difficult for the air to penetrate the layer and an inner anaerobic layer is formed.

The components of a complete trickling filter system are: fundamental components:

- A bed of filter medium upon which a layer of microbial slime is promoted and developed.
- An enclosure or a container which houses the bed of filter medium.

- A system for distributing the flow of wastewater over the filter medium.
- A system for removing and disposing of any sludge from the treated effluent.

The treatment of sewage or other wastewater with trickling filters is among the oldest and most well characterized treatment technologies.

A trickling filter is also often called a *trickle filter, trickling biofilter, biofilter, biological filter* or *biological trickling filter*.

Treatment of Other Organics

Synthetic organic materials including solvents, paints, pharmaceuticals, pesticides, coking products and so forth can be very difficult to treat. Treatment methods are often specific to the material being treated. Methods include Advanced Oxidation Processing, distillation, adsorption, vitrification, incineration, chemical immobilisation or landfill disposal. Some materials such as some detergents may be capable of biological degradation and in such cases, a modified form of wastewater treatment can be used.

Treatment of Acids and Alkalis

Acids and alkalis can usually be neutralised under controlled conditions. Neutralisation frequently produces a precipitate that will require treatment as a solid residue that may also be toxic. In some cases, gasses may be evolved requiring treatment for the gas stream. Some other forms of treatment are usually required following neutralisation.

Waste streams rich in hardness ions as from de-ionisation processes can readily lose the hardness ions in a buildup of precipitated calcium and magnesium salts. This precipitation process can cause severe *furring* of pipes and can, in extreme cases, cause the blockage of disposal pipes. A 1 metre diameter industrial marine discharge pipe serving a major chemicals complex was blocked by such salts in the 1970s. Treatment is by concentration of de-ionisation waste waters and disposal to landfill or by careful pH management of the released wastewater.

Treatment of Toxic Materials

Toxic materials including many organic materials, metals (such as zinc, silver, cadmium, thallium, etc.) acids, alkalis, non-metallic elements (such as arsenic or selenium) are generally resistant to biological processes unless very dilute. Metals can often be precipitated out by changing the pH or by treatment with other chemicals. Many, however, are resistant to treatment or mitigation and may require concentration followed by landfilling or recycling. Dissolved organics can be *incinerated* within the wastewater by Advanced Oxidation Process.

Agricultural Wastewater

Nonpoint Source Controls: Sediment (loose soil) washed off fields is the largest source of agricultural pollution in the United States. Farmers may utilize erosion controls to reduce runoff flows and retain soil on their fields. Common techniques include contour plowing, crop mulching, crop rotation, planting perennial crops and installing riparian buffers. Nutrients (nitrogen and phosphorus) are typically applied to farmland as commercial fertilizer; animal manure; or spraying of municipal or industrial wastewater (effluent) or sludge. Nutrients may also enter runoff from crop residues, irrigation water, wildlife, and atmospheric deposition. Farmers can develop and implement nutrient management plans to reduce excess application of nutrients. To minimize pesticide impacts, farmers may use Integrated Pest Management (IPM) techniques (which can include biological pest control) to maintain control over pests, reduce reliance on chemical pesticides, and protect water quality.

Point Source Wastewater Treatment: Farms with large livestock and poultry operations, such as factory farms, are called *concentrated animal feeding operations* or *feedlots* in the US and are being subject to increasing government regulation. Animal slurries are usually treated by containment in anaerobic lagoons before disposal by spray or trickle application to grassland. Constructed wetlands are sometimes used to facilitate treatment of animal wastes. Some animal slurries are treated by mixing with straw and composted at high temperature to produce a bacteriologically sterile and friable manure for soil improvement.

Construction Site Stormwater

Sediment from construction sites is managed by installation of:

- erosion controls, such as mulching and hydroseeding, and
- sediment controls, such as sediment basins and silt fences.

Discharge of toxic chemicals such as motor fuels and concrete washout is prevented by use of:

- spill prevention and control plans, and

- specially designed containers (e.g. for concrete washout) and structures such as overflow controls and diversion berms.

Urban Runoff (Stormwater)

Urban runoff is surface runoff of rainwater created by urbanization. This runoff is a major source of water pollution in urban communities worldwide. Impervious surfaces (roads, parking lots and sidewalks) are constructed during land development. During rain storms and other precipitation events, these surfaces (built from materials such as asphalt and concrete), along with rooftops, carry polluted stormwater to storm drains, instead of allowing the water to percolate through soil. This causes lowering of the water table (because groundwater recharge is lessened) and flooding since the amount of water that remains on the surface is greater. Most municipal storm sewer systems discharge stormwater, untreated, to streams, rivers and bays.

Pollutants in Urban Runoff

Water running off these impervious surfaces tends to pick up gasoline, motor oil, heavy metals, trash and other pollutants from roadways and parking lots, as well as fertilizers and pesticides from lawns. Roads and parking lots are major sources of polycyclic aromatic hydrocarbons (PAHs), which are created as combustion byproducts of gasoline and other fossil fuels, as well as of the heavy metals nickel, copper, zinc, cadmium, and lead. Roof runoff contributes high levels of synthetic organic compounds and zinc (from galvanized gutters). Fertilizer use on residential lawns, parks and golf courses is a significant source of nitrates and phosphorus in urban runoff.

As stormwater is channeled into storm drains and surface waters, the natural sediment load discharged to receiving waters decreases, but the water flow and velocity increases. In fact, the impervious cover in a typical city creates five times the runoff of a typical woodland of the same size.

Effects of Urban Runoff

A 2008 report by the United States National Research Council (textbox below) identified urban runoff as a leading source of water quality problems. Further declines in water quality remain likely if the land-use changes that typify more diffuse sources of pollution are not addressed... These include land-disturbing agricultural, silvicultural, urban, industrial, and construction activities from which hard-to-monitor pollutants emerge during wet-weather events. Pollution from these landscapes has been almost universally acknowledged as the most pressing challenge to the restoration of waterbodies and aquatic ecosystems nationwide.

The runoff also increases temperatures in streams, harming fish and other organisms. (A sudden burst of runoff from a rainstorm can cause a fish-killing shock of hot water.) Also, road salt used to melt snow on sidewalks and roadways can contaminate streams and groundwater aquifers.

One of the most pronounced effects of urban runoff is on watercourses that historically contained little or no water during dry weather periods (often called *ephemeral streams*). When an area around such a stream is urbanized, the resultant runoff creates an unnatural year-round streamflow that hurts the vegetation, wildlife and stream bed of the waterway. Containing little or no sediment relative to the historic ratio of sediment to water, urban runoff rushes down the stream channel, ruining natural features such as meanders and sandbars, and creates severe erosion—increasing sediment loads at the mouth while severely incising the stream bed upstream. As an example, on many Southern California beaches at the mouth of a waterway, urban runoff carries trash, pollutants, excessive silt, and other wastes, and can pose moderate to severe health hazards.

Because of fertilizer and organic waste that urban runoff often carries, eutrophication often occurs in waterways affected by this type of runoff. After heavy rains, organic matter in the waterway is relatively high compared with natural levels, spurring growth of algae blooms that soon use up most of the oxygen. Once the naturally occurring oxygen in the water is depleted, the algae blooms die, and in their decomposing process, cause further eutrophication. Algae blooms mostly occur in areas with still water, such as stream pools and the pools behind dams, weirs, and some drop structures. Eutrophication usually comes with deadly consequences for fish and other aquatic organisms. Excessive stream bank erosion may cause flooding and property damage. For many years governments have often responded to urban stream erosion problems by modifying the streams through construction of hardened embankments and similar control structures using concrete and masonry materials. Use of these hard materials destroys habitat for fish and other animals. Such a project may stabilize the immediate area where flood damage occurred, but often it simply shifts the problem to an upstream or downstream segment of the stream.

Prevention and Mitigation of Urban Runoff

Effective control of urban runoff involves reducing the velocity and flow of stormwater, as well as reducing pollutant discharges. A variety of stormwater management practices and systems may be used to reduce the effects of urban runoff. Some of these techniques (called best management practices (BMPs)) in the US, focus on water quantity control, while others focus on improving water quality, and some perform both functions. Pollution prevention practices include low impact development techniques, installation of green roofs and improved chemical handling (e.g. management of motor fuels & oil, fertilizers and pesticides). Runoff mitigation systems include infiltration basins, bioretention systems, constructed wetlands, retention basins and similar devices.

Effective control of urban runoff involves reducing the velocity and flow of stormwater, as well as reducing pollutant discharges. Local governments use a variety of stormwater management techniques to reduce the effects of urban runoff. These techniques, called best management practices (BMPs) in the U.S., may focus on water quantity control, while others focus on improving water quality, and some perform both functions. Pollution prevention practices include low-impact development techniques, installation of green roofs and improved chemical handling (e.g. management of motor fuels & oil, fertilizers and pesticides). Runoff mitigation systems include infiltration basins, bioretention systems, constructed wetlands, retention basins and similar devices.

Thermal pollution from runoff can be controlled by stormwater management facilities that absorb the runoff or direct it into groundwater, such as bioretention systems and infiltration basins. Retention basins tend to be less effective at reducing temperature, as the water may be heated by the sun before being discharged to a receiving stream.Rainwater Harvesting

Rainwater harvesting is the accumulating and storing of rainwater for reuse before it reaches the aquifer. It has been used to provide drinking water, water for livestock, water for irrigation, as well as other typical uses. Rainwater collected from the roofs of houses and local institutions can make an important contribution to the availability of drinking water. It can supplement the subsoil water level and increase urban greenery. Water collected from the ground, sometimes from areas that are especially prepared for this purpose, is called Stormwater harvesting. In some cases, rainwater may be the only available, or economical, water source. Rainwater harvesting systems can be simple to construct from inexpensive local materials, and are potentially successful in most habitable locations. Roof rainwater may not be potable and may require treatment before consumption. As rainwater rushes from your roof it may carry pollutants, such as mercury from coal burning buildings, or bird faeces. Although some rooftop materials may produce rainwater that would be harmful to human health as drinking water, it can be useful in flushing toilets, washing clothes, watering the garden, and washing cars; these uses alone halve the amount of water used by a typical home. Household rainfall catchment systems are appropriate in areas with an average rainfall greater than 200 mm (7.9 in) per year, and no other accessible water sources (Skinner and Cotton, 1992). Overflow from rainwater harvesting tank systems can be used to refill aquifers in a process called groundwater recharge; though this is a related process, it must not be confused with rainwater harvesting.

There are several types of systems to harvest rainwater, ranging from very simple home systems to complex industrial systems. The rate at which water can be collected from either system is dependent on the plan area of the system, its efficiency, and the intensity of rainfall (i.e., annual precipitation (mm per annum) x square meter of catchment area = litres per annum yield) ... a 200 square meter roof catchment catching 1,000mm PA yields 200 kLPA.

Basic configuration

Rainwater harvesting systems channel rainwater from a roof into storage via an arrangement of gutters and pipes. The first flush of rainwater after a dry season should be allowed to run to waste, as it will be contaminated with dust, bird droppings etc. Roof gutters should be large enough to carry peak flows. Storage tanks should be covered to prevent mosquito breeding and to reduce evaporation losses, contamination and algal growth. Rainwater harvesting systems require regular maintenance and cleaning to keep the system hygienic.

Subsurface Dike

A subsurface dike is built in an aquifer to obstruct the natural flow of groundwater, thereby raising the groundwater level and increasing the amount of water stored in the aquifer. The subsurface dike at Krishi Vigyan Kendra, Kannur under Kerala Agricultural University with the support of ICAR, has become an effective method for ground water conservation by means of rain water harvesting technologies. The subsurface dike has been demonstrated to be a feasible method for conserving and exploiting the groundwater resources of the Kerala state of India. The dike is now the largest rainwater harvesting system in that region.

Groundwater Recharge

Rainwater may also be used for groundwater recharge, where the runoff on the ground is collected and allowed to be absorbed, adding to the groundwater. In the US, rooftop rainwater is collected and stored in sump.

Advantages in Urban Areas

Rainwater harvesting can ensure an independent water supply during water restrictions, though somewhat dependent on end-use and maintenance, usually of acceptable quality for household needs and renewable at acceptable volumes, despite forecasted climate change (CSIRO, 2003). It produces beneficial externalities by reducing peak storm water runoff and processing costs. In municipalities with combined sewer systems, reducing storm runoff is especially important, because excess runoff during heavy storms leads to the discharge of raw sewage from outfalls when treatment plant capacity cannot handle the combined flow. Rainwater harvesting systems are simple to install and operate. Running costs are negligible, and they provide water at the point of consumption. Rainwater harvesting in urban communities has been made possible by various companies. Their tanks provide an attractive yet effective solution to rainwater catchment.

Rain water harvesting law Some U.S. states have water law in favor of land owners, and some states own all water rights, leaving the owner only water rights the states are willing to grant through permits. For example, in Colorado you may not catch, collect or harvest rainwater from your roof unless you first buy a permit.

In Australia In Australia it is common to have a rain water tank with rainwater running off the house roof to fill tank, it is common to flush toilets with this water and have a back up valve like an Acquasaver that gives the home owner mains(town water) as a back up in case rainwater supply is exhausted or power failure. These types of valves automatically switch from rain water to mains water supply when available, so the homeowner never needs manually operate this valve.

Quality

As rainwater may be contaminated due to pollutants like microscopic germs etc., it is often not considered suitable for drinking without treatment. However, there are many examples of rainwater being used for all purposes — including drinking — following suitable treatment.

Rainwater harvested from roofs can contain human, animal and bird faeces, mosses and lichens, windblown dust, particulates from urban pollution, pesticides, and inorganic ions from the sea (Ca, Mg, Na, K, Cl, SO_4), and dissolved gases (CO_2, NO_x, SO_x). High levels of pesticide have been found in rainwater in Europe with the highest concentrations occurring in the first rain immediately after a dry spell; the concentration of these and other contaminants are reduced significantly by diverting the initial flow of water to waste as described above. The water may need to be analysed properly, and used in a way appropriate to its safety. In the Gansu province for example, harvested rainwater is boiled in parabolic solar cookers before being used for drinking. In Brazil alum and chlorine is added to disinfect water before consumption. So-called "appropriate technology" methods, such as solar water disinfection, provide low-cost disinfection options for treatment of stored rainwater for drinking.

System Sizing

It is important that the system is sized to meet the water demand throughout the dry season. In general, the size of the storage tank should be big enough to meet the daily water requirement throughout the dry season. In addition, the size of the catchment area or roof should be large enough to fill the tank.

Around the World

- Currently in China and Brazil, rooftop rainwater harvesting is being practiced for providing drinking water, domestic water, water for livestock, water for small irrigation and a way to replenish ground water levels. Gansu province in China and semi-arid north east Brazil have the largest rooftop rainwater harvesting projects ongoing.
- In Tamil Nadu, India rainwater harvesting was made compulsory for every building to avoid ground water depletion. It proved excellent results within five years and every other state took it as role model. Since the implementation, Chennai saw 50 per cent rise in water level in five years and the water quality significantly improved.
- In Rajasthan, India rainwater harvesting has traditionally been practiced by the people of the Thar Desert. There are many ancient water harvesting systems in Rajasthan, which have now been revived
- In Bermuda, the law requires all new construction to include rainwater harvesting adequate for the residents.

- The U.S. Virgin Islands have a similar law.
- In the Indus Valley Civilization, Elephanta Caves and Kanheri Caves in Mumbai rainwater harvesting alone has been used to supply in their water requirements.
- In Senegal and Guinea-Bissau, the houses of the Diola-people are frequently equipped with homebrew rainwater harvesters made from local, organic materials.
- In the United Kingdom water butts are often found in domestic gardens to collect rainwater, which is then used to water the garden. However, the British government's Code For Sustainable Homes encourages fitting large underground tanks to new-build homes to collect rainwater for flushing toilets, washing clothes, watering the garden, and washing cars. This reduces by 50% the amount of mains water used by the home.
- In the Irrawaddy Delta of Myanmar, the groundwater is saline and communities rely on mud-lined rainwater ponds to meet their drinking water needs throughout the dry season. Some of these ponds are centuries old and are treated with great reverence and respect.
- Until 2009 in Colorado, water rights laws almost completely restricted rainwater harvesting; a property owner who captured rainwater was deemed to be stealing it from those who have rights to take water from the watershed. Now, residential well owners that meet certain criteria may obtain a permit to install a *rooftop precipitation collection system* (SB 09-080). Up to 10 large scale pilot studies may also be permitted (HB 09-1129). The main factor in persuading the Colorado Legislature to change the law was a 2007 study that found that in an average year, 97% of the precipitation that fell in Douglas County, in the southern suburbs of Denver, never reached a stream—it was used by plants or evaporated on the ground. In colorado you cannot even drill a water well unless you have at least 35 acres. In New Mexico, rainwater catchment is mandatory for new dwellings in Santa Fe.
- Kerala, India,
- In Australia rainwater harvesting is typically used to supplement the reticulated mains supply. In south east Queensland, households that harvested rainwater doubled each year from 2005 to 2008, reaching 40% penetration at that time (White, 2009 (PhD)).

- In Beijing, some housing societies are now adding rain water in their main water sources after proper treatment.

Professor Micheal McGinley established a project to design a rain water harvesting prototype in the Biosystems design Challenge Module in University College Dublin.

Green Infrastructure

Green Infrastructure is a concept originating in the United States in the mid-1990s that highlights the importance of the natural environment in decisions about land-use planning. In particular there is an emphasis on the "life support" functions provided by a network of natural ecosystems, with an emphasis on interconnectivity to support long-term sustainability. Examples include clean water and healthy soils, as well as the more anthropocentric functions such as recreation and providing shade and shelter in and around towns and cities.

The United States Environmental Protection Agencyý (EPA) has extended the concept to apply to the management of stormwater runoff at the local level through the use of natural systems, or engineered systems that mimic natural systems, to treat polluted runoff. This use of the term "green infrastructure" to refer to urban "green" best management practices (BMPs), although not central to the larger concept, does contribute to the overall health of natural ecosystems.

Planning Approach

The Green Infrastructure approach analyses the natural environment in a way that highlights its function and subsequently seeks to put in place, through regulatory or planning policy, mechanisms that safeguard critical natural areas. Where life support functions are found to be lacking, plans may propose how these can be put in place through landscaped and/or engineered improvements. The term "green infrastructure" is sometimes expanded to "multifunctional" green infrastructure. Multifunctionality in this context refers to the integration and interaction of different functions or activities on the same piece of land. This is key to the efficient and sustainable use of land, especially in a compact and bustling country like England where pressures on land are particularly acute. An example might be an urban edge river floodplain which provides a repository for flood waters, acts as a nature reserve, provides a recreational green space and could also be productively farmed (probably through grazing). There is growing evidence that the natural environment also has a positive effect on human health.

UK Applications

In the United Kingdom, Green Infrastructure planning is increasingly recognised as a valuable approach for spatial planning and is now seen in national, regional and local planning and policy documents and strategies, for example in the Milton Keynes and South Midlands Growth area.

In 2009, guidance on green infrastructure planning was published by Natural England, and can be accessed online at [2] This guidance promotes the importance of green infrastructure in 'place-making', i.e. in recognising and maintaining the character of a particular location, especially where new development is planned.

In North West England the former *Regional Spatial Strategy* had a specific Green Infrastructure Policy (EM3 - Green Infrastructure) as well as other references to the concept in other land use development policies (e.g. DP6). The policy was supported by the *North West Green Infrastructure Guide.* The Green Infrastructure Think Tank (GrITT) provides the support for policy development in the region and manages the web site that acts as a repository for information on Green Infrastructure.

The Natural Economy Northwest programme has supported a number of projects, commissioned by The Mersey Forest to develop the evidence base for green infrastructure in the region. In particular it has work has been undertaken to look at the economic value of green infrastructure, the linkage between grey and green infrastructure and also to identify areas where green infrastructure may play critical role in helping to overcome issues such as risks of flood or poor air quality.

In March 2011, a prototype Green Infrastructure Valuation Toolkit was launched. The Toolkit is available under a Creative Commons licence, and provides a range of tools that provide economic valuation of green infrastructure interventions. The toolkit has been trialled in a number of areas and strategies, including the Liverpool Green Infrastructure Strategy.

Green Infrastructure is being promoted as an effective and efficient response to projected climate change.

US Applications

Green infrastructure programs managed by EPA and partner organizations are intended to improve water quality generally through more extensive management of stormwater runoff. The practices are expected to reduce stress on traditional water drainage infrastructure—storm sewers and combined sewers—which are typically extensive networks of underground pipes and/or surface water channels in U.S. cities, towns and suburban areas. Improved stormwater management is expected to reduce the frequency of combined sewer overflows and sanitary sewer overflows, and provide other environmental benefits.

For example, the City of Philadelphia has installed or supported a variety of retrofit projects in neighborhoods throughout the city. Installed improvements include:

- permeable pavements in parks, basketball courts and parking lots
- rain gardens and bioretention systems at schools and other public facilities
- constructed wetlands for management of stormwater runoff.

Some of these facilities reduce the volume of runoff entering the city's aging combined sewer system, and thereby reduce the extent of system overflows during rainstorms.

Another US example, the State of Maryland, has promoted a program called GreenPrint. GreenPrint Maryland is a first in the nation web-enabled map showing the relative ecological importance of every parcel of land in the State.

Combining color-coded maps, information layers, and aerial photography with public openness and transparency, Greenprint Maryland applies the best environmental science and Geographic Information Systems (GIS) to the urgent work of preserving and protecting environmentally critical lands. A valuable new tool not only for making land conservation decisions today, but for building a broader and better informed public consensus for sustainable growth and land preservation decisions into the future.

The program was established in 2001 with the objective to "preserve an extensive, intertwined network of land vital to long term survival of our native plants and wildlife and industries dependent on clean environment and abundant natural resources."

Sanitation

Sanitation is the hygienic means of promoting health through prevention of human contact with the hazards of wastes. Hazards can be either physical, microbiological, biological or chemical agents of disease. Wastes that can cause health problems are human and animal feces, solid wastes, domestic wastewater (sewage, sullage, greywater), industrial wastes and agricultural wastes. Hygienic means of prevention can be by using engineering solutions (e.g. sewerage and wastewater treatment), simple technologies (e.g. latrines, septic tanks), or even by personal hygiene practices (e.g. simple handwashing with soap).

Overview

The World Health Organization States that: "Sanitation generally refers to the provision of facilities and services for the safe disposal of human urine and faeces. Inadequate sanitation is a major cause of disease world-wide and improving sanitation is known to have a significant beneficial impact on health both in households and across communities. The word 'sanitation' also refers to the maintenance of hygienic conditions, through services such as garbage collection and wastewater disposal.

The term "sanitation" can be applied to a specific aspect, concept, location or strategy, such as:

- Basic sanitation - refers to the management of human faeces at the household level. This terminology is the indicator used to describe the target of the Millennium Development Goal on sanitation.
- On-site sanitation - the collection and treatment of waste is done where it is deposited. Examples are the use of pit latrines, septic tanks, and Imhoff tanks.
- Food sanitation - refers to the hygienic measures for ensuring food safety.
- Environmental sanitation - the control of environmental factors that form links in disease transmission. Subsets of this category are solid waste management, water and wastewater treatment, industrial waste treatment and noise and pollution control.
- Ecological sanitation - an approach that tries to emulate nature through the recycling of nutrients and water from human and animal wastes in a hygienically safe manner.

History

The earliest evidence of urban sanitation was seen in Harappa, Mohenjo-daro and the recently discovered Rakhigarhi of Indus Valley civilization. This urban plan included the world's first urban sanitation systems. Within the city, individual homes or groups of homes obtained water from wells. From a room that appears to have been set aside for bathing, waste water was directed to covered drains, which lined the major streets.

Roman cities and Roman villas had elements of sanitation systems, delivering water in the streets of towns such as Pompeii, and building stone and wooden drains to collect and remove wastewater from populated areas. But there is little record of other sanitation in most of Europe until the High Middle Ages. Unsanitary conditions and overcrowding were widespread throughout Europe and Asia during the Middle Ages, resulting periodically in cataclysmic pandemics such as the Plague of Justinian (541-42) and the Black Death (1347–1351), which killed tens of millions of people and radically altered societies.

Very high infant and child mortality prevailed in Europe throughout medieval times, due not only to deficiencies in sanitation but to an insufficient food supply for a population which had expanded faster than agriculture. This was further complicated by frequent warfare and exploitation of civilians by autocratic rulers.

Wastewater Sanitation

Wastewater Collection: The standard sanitation technology in urban areas is the collection of wastewater in sewers, its treatment in wastewater treatment plants for reuse or disposal in rivers, lakes or the sea. Sewers are either combined with storm drains or separated from them as sanitary sewers. Combined sewers are usually found in the central, older parts or urban areas. Heavy rainfall and inadequate maintenance can lead to combined sewer overflows or sanitary sewer overflows, i.e. more or less diluted raw sewage being discharged into the environment. Industries often discharge wastewater into municipal sewers, which can complicate wastewater treatment unless industries pre-treat their discharges.

The high investment cost of conventional wastewater collection systems are difficult to afford for many developing countries. Some countries have therefore promoted alternative wastewater collection systems such as condominial sewerage, which uses smaller diameter pipes at lower depth with different network layouts from conventional sewerage.

Wastewater Treatment

In developed countries treatment of municipal wastewater is now widespread, but not yet universal. In developing countries most wastewater is still discharged untreated into the environment. For example, in Latin America only about 15% of collected sewerage is being treated.

Reuse of Wastewater

The reuse of untreated wastewater in irrigated agriculture is common in developing countries. The reuse of treated wastewater in landscaping, especially on golf courses, irrigated agriculture and for industrial use is becoming increasingly widespread.

In many suburban and rural areas households are not connected to sewers. They discharge their wastewater into septic tanks or other types of on-site sanitation.

Ecological Sanitation

Ecological sanitation is sometimes presented as a radical alternative to conventional sanitation systems. Ecological sanitation is based on composting or vermicomposting toilets where an extra separation of urine and feces at the source for sanitization and recycling has been done. It thus eliminates the creation of blackwater and eliminates fecal pathogens. If ecological sanitation is practiced municipal wastewater consists only of greywater, which can be recycled for gardening. However, in most cases greywater continues to be discharged to sewers.

An ecological sanitation (ecosan) viewpoint sees human waste and wastewater as an opportunity. When properly designed and operated, ecosan systems provide a hygienically safe, economical, and closed-loop system to convert human wastes into nutrients to be returned to the soil, and water to be returned to the land. Alternatively, solid wastes are converted into a biofuel. The primary application for ecosan systems has been in rural areas where connection to a sanitary sewer system is not possible, or where water supplies are very limited.

The main objectives of ecological sanitation are:

- To reduce the health risks related to sanitation, contaminated water and waste
- To prevent the pollution of surface and ground water
- To reuse nutrients or energy contained within wastes.

History of Reuse-oriented Sanitation Approaches

The recovery and use of urine and feces has been practiced by almost all cultures. The reuse was not limited to agricultural production. The Romans, for example, were aware of the bleaching attribute of the ammonia within urine and used it to whiten clothing.

The most widely known reuse in agriculture has occurred in China. Reportedly, the Chinese were aware of the benefits of using excreta in crop production before 500 B.C., enabling them to sustain more people at a higher density than any other system of agriculture. The value of "night soil" as a fertilizer was recognized with well-developed systems in place to enable the collection of excreta from cities and its transportation to fields. However, its use promoted disease to such an extent that in Chinese cuisine almost all vegetables are thoroughly cooked. Elaborate systems were developed in urban centers of Yemen enabling the separation of urine and excreta even in multi-story buildings. Feces were collected from toilets via vertical drop shafts, while urine did not enter the shaft but passed instead along a channel leading through the wall to the outside where it evaporated. Here, feces were not used in agriculture but were dried and burnt as a biofuel.

In Mexico and Peru, both the Aztec and Inca cultures collected human excreta for agricultural use. In Peru, the Incas had a high regard for excreta as a fertilizer, which was stored, dried and pulverized to be utilized when planting maize.

In the Middle Ages, the use of excreta and greywater was the norm. European cities were rapidly urbanizing and sanitation was becoming an increasingly serious problem, whilst at the same time the cities themselves were becoming an increasingly important source of agricultural nutrients. The practice of using the nutrients in excreta and wastewater for agriculture therefore continued in Europe into the middle of the 19th Century. Farmers, recognizing the value of excreta, were eager to get these fertilizers to increase production and urban sanitation benefited.

The increasing number of research and demonstration projects for excreta reuse carried out in Sweden from the 1980s to the early 21st century aimed at developing hygienically safe closed loop sanitation systems. Similar lines of research began elsewhere, for example in Zimbabwe, in the Netherlands, Norway and Germany. These closed-loop sanitation systems became popular under the name "ecosan", "dewats", "desar", and other abbreviations. They placed their emphasis on the hygenisation of the contaminated flow streams, and shifted the concept from waste disposal to resource conservation and safe reuse.

Concepts of Ecological Sanitation

Ecological sanitation (Ecosan) is based on an overall view of material flows as part of an ecologically and economically sustainable wastewater management system tailored to the needs of the users and to the respective local conditions. It does not favour a specific sanitation technology, but is rather a new philosophy in handling substances that have so far been seen simply as wastewater and water-carried waste for disposal.

According to Esrey et al. (2003) ecological sanitation can be defined as a system that:

- Prevents disease and promotes health
- Protects the environment and conserves water
- Recovers and recycles nutrients and organic matter

Ecosan offers a flexible framework, where centralised elements can be combined with decentralised ones, waterborne with dry sanitation, high-tech with low-tech, etc. By considering a much larger range of options, optimal and economic solutions can be developed for each particular situation.

Thus, the most important advantages of ecological sanitation systems are:

- Improvement of health by minimising the introduction of pathogens from human excreta into the water cycle
- Promotion of safe, hygienic recovery and use of nutrients, organics, trace elements, water and energy
- Preservation of soil fertility
- Contribution to the conservation of resources through lower water consumption, substitution of mineral fertiliser and minimisation of water pollution
- Improvement of agricultural productivity and food security
- Preference for modular, decentralised partial-flow systems for more appropriate cost-efficient solutions adapted to the local situation
- Promotion of a holistic, interdisciplinary approach
- Material flow cycle instead of disposal of valuable resources

Project Examples

Examples of ecosan projects can be found among others in the collection of project data sheets of gtz ecosan or on the Enhanced Global Map of ecosan activities by EcoSanRes. In the following some examples are given that underline the diversity of ecosan projects:

Guangxi province, China - large-scale project of urine diverting dehydration toilets The dissemination programme of ecological dry toilets for Hsinchu County, Guangxi province, one of the poorest provinces in China, started in 1997 with support of UNICEF, SIDA and the Red Cross and has been expanded to 17 provinces until the year 2003. By this year, the scale of the project had increased to approximately 685,000 toilet units – today more than one million double vault urine diversion dehydration toilets (UDDTs) are installed in rural areas of China.

In UDDTs, urine and faeces are collected separately: The urine is collected in the front and led by a plastic pipe to a storage canister from where it can be used as a fertilizer in agriculture, the faeces fall at the back in one of two ventilated storage chambers and are covered with ash for better dehydration. After about one year of storage the dried material can be removed and used as a soil conditioner in agriculture.

KfW, Frankfurt, Germany - vacuum toilets + greywater treatment The sanitation concept of the modern office building "Ostarkarde" of the KfW Bankengruppe in Frankfurt is based on a separate excreta and greywater collection. While urine and faeces are collected via vacuum toilets and a vacuum sewerage using much less water for flushing, the greywater from hand washing and kitchen is collected and treated separately in a compact activated sludge reactor combined with membrane filtration. The treated greywater is then reused for toilet flushing and cleaning water. The amount of greywater can be reduced by 76% by this cost-efficient system which could be one of the prior choices for sanitation systems of newly constructed office buildings.

Tanum Municipality in Sweden has introduced urine separation toilets to recover phosphorus.

Arguments for the Use of Ecological Sanitation

Often, water used in flush toilets is of drinking quality. Only 1% of global water is drinkable, therefore, it is a precious resource. Water fit to drink is being used for other purposes that can use lesser quality water, such as toilets.

Mixing feces and urine makes treatment difficult. All waste water treatment plants use some natural/biological processes, but nature does not normally have this waste water, so there are no microbes that can deal with this mix. In order to treat waste, treatment plants have to do this in stages. Each stage treats a different component of the mix by creating the right environment for microbes to do their work (aerobic, anaerobic, anoxic and the right pH). This is costly and requires energy.

A mix of domestic and industrial effluent in water cannot be treated properly, for heavy metals and other pollutants make this water unsuitable for reuse. This is normally discharged into the ground or water bodies.

Because of the complexity of the treatment process, treatment plants tend to be large. This requires costly infrastructure to build and maintain it, often out of the reach of poorer communities.

John Jeavons argues that "Each person's urine and manure contain approximately enough nutrients to produce enough food to feed that person." Urea is the major component of urine, yet we produce vast quantities of urea by using fossil fuels. By properly managing urine, treatment costs as well as fertilizer costs can be reduced. Feces also contains recognized nutrients, and could be used for modern agriculture, as micronutrient deficiency is a significant problem.

Sanitation and Public Health

The importance of the isolation of waste lies in an effort to prevent diseases which can be transmitted through human waste, which afflict both developed countries as well as developing countries to differing degrees. It is estimated that up to 5 million people die each year from preventable water-borne disease, as a result of inadequate sanitation and hygiene practices. The affects of sanitation have also had a large impact on society. The results of studies published in *Griffins Public Sanitation* show that better sanitation produces an enhanced feeling of wellbeing.

Relevant disease include:

- Waterborne diseases, which can contaminate drinking water
- Diseases transmitted by the fecal-oral route
- Hookwork, where eggs can survive in the soil

Global Access to Improved Sanitation

The Joint Monitoring Program for water and sanitation of WHO and UNICEF has defined improved sanitation as

- connection to a public sewer
- connection to a septic system
- pour-flush latrine

- simple pit latrine
- ventilated improved pit latrine

According to that definition, 62% of the world's population has access to improved sanitation in 2008, up by 8% since 1990. Only slightly more than half of them or 31% of the world population lived in houses connected to a sewer. Overall, 2.5 billion people lack access to improved sanitation and thus must resort to open defecation or other unsanitary forms of defecation, such as public latrines or open pit latrines. This includes 1.2 billion people who have access to no facilities at all. This outcome presents substantial public health risks as the waste could contaminate drinking water and cause life threatening forms of diarrhea to infants. Improved sanitation, including hand washing and water purification, could save the lives of 1.5 million children who die from diarrheal diseases each year.

In developed countries, where less than 20% of the world population lives, 99% of the population has access to improved sanitation and 81% were connected to sewers.

Solid Waste Disposal

Disposal of solid waste is most commonly conducted in landfills, but incineration, recycling, composting and conversion to biofuels are also avenues. In the case of landfills, advanced countries typically have rigid protocols for daily cover with topsoil, where underdeveloped countries customarily rely upon less stringent protocols. The importance of daily cover lies in the reduction of vector contact and spreading of pathogens. Daily cover also minimises odor emissions and reduces windblown litter. Likewise, developed countries typically have requirements for perimeter sealing of the landfill with clay-type soils to minimize migration of leachate that could contaminate groundwater (and hence jeopardize some drinking water supplies).

For incineration options, the release of air pollutants, including certain toxic components is an attendant adverse outcome. Recycling and biofuel conversion are the sustainable options that generally have superior life cycle costs, particularly when total ecological consequences are considered. Composting value will ultimately be limited by the market demand for compost product.

Sanitation in the Developing World

The United Nations Millennium Development Goals (MDGs) include a target to reduce by half the proportion of people without access to basic sanitation by 2015. In December 2006, the United

Nations General Assembly declared 2008 'The International Year of Sanitation', in recognition of the slow progress being made towards the MDGs sanitation target. The year aims to develop awareness and action to meet the target. Particular concerns are:

- Removing the stigma around sanitation, so that the importance of sanitation can be more easily and publicly discussed.
- Highlighting the poverty reduction, health and other benefits that flow from better hygiene, household sanitation arrangements and wastewater treatment.

Research from the Overseas Development Institute suggests that sanitation and hygiene promotion needs to be better 'mainstreamed' in development, if the MDG on sanitation is to be met. At present, promotion of sanitation and hygiene is mainly carried out through water institutions. The research argues that there are, in fact, many institutions that should carry out activities to develop better sanitation and hygiene in developing countries. For example, educational institutions can teach on hygiene, and health institutions can dedicate resources to preventative works (to avoid, for example, outbreaks of cholera).

The Institute of Development Studies (IDS) coordinated research programme on Community-led Total Sanitation (CLTS) is a radically different approach to rural sanitation in developing countries and has shown promising successes where traditional rural sanitation programmes have failed. CLTS is an unsubsidized approach to rural sanitation that facilitates communities to recognize the problem of open defecation and take collective action to clean up and become 'open defecation free'. It uses community-led methods such as participatory mapping and analysing pathways between feces and mouth as a means of galvanizing communities into action. An IDS 'In Focus' Policy Brief suggests that in many countries the Millennium development goal for sanitation is off track and asks how CLTS can be adopted and spread on a large scale in the many countries and regions where open defecation still prevails.

Sanitation in the Food Industry

Sanitation within the food industry means the adequate treatment of food-contact surfaces by a process that is effective in destroying vegetative cells of microorganisms of public health significance, and in substantially reducing numbers of other undesirable microorganisms, but without adversely affecting the food or its safety for the consumer (U.S. Food and Drug Administration, Code of Federal Regulations,

21CFR110, USA). Sanitation Standard Operating Procedures are mandatory for food industries in US, which are regulated by 9 CFR part 416 in conjunction with 21 CFR part 178.1010. Similarly, in Japan, food hygiene has to be achieved through compliance with food sanitation law.

Additionally, in the food and biopharmaceutical industries, the term sanitary equipment means equipment that is fully cleanable using clean-in-place (CIP) and sterilization-in-place (SIP) procedures: that is fully drainable from cleaning solutions and other liquids. The design should have a minimum amount of deadleg or areas where the turbulence during cleaning is insufficient to remove product deposits. In general, to improve cleanability, this equipment is made from Stainless Steel 316L, (an alloy containing small amounts of molybdenum). The surface is usually electropolished to an effective surface roughness of less than 0.5 micrometre to reduce the possibility of bacterial adhesion.

Bio-gas Plant

Biogas typically refers to a gas produced by the biological breakdown of organic matter in the absence of oxygen. Organic waste such as dead plant and animal material, animal dung, and kitchen waste can be converted into a gaseous fuel called biogas. Biogas originates from biogenic material and is a type of biofuel.

Biogas is produced by the anaerobic digestion or fermentation of biodegradable materials such as biomass, manure, sewage, municipal waste, green waste, plant material, and crops. Biogas comprises primarily methane (CH_4) and carbon dioxide (CO_2) and may have small amounts of hydrogen sulphide (H_2S), moisture and siloxanes.

The gases methane, hydrogen, and carbon monoxide (CO) can be combusted or oxidized with oxygen. This energy release allows biogas to be used as a fuel. Biogas can be used as a fuel in any country for any heating purpose, such as cooking. It can also be used in anaerobic digesters where it is typically used in a gas engine to convert the energy in the gas into electricity and heat. Biogas can be compressed, much like natural gas, and used to power motor vehicles. In the UK, for example, biogas is estimated to have the potential to replace around 17% of vehicle fuel. Biogas is a renewable fuel, so it qualifies for renewable energy subsidies in some parts of the world. Biogas can also be cleaned and upgraded to natural gas standards when it becomes biomethane.

Production

Biogas is practically produced as landfill gas (LFG) or digester gas. A *biogas plant* is the name often given to an anaerobic digester that treats farm wastes or energy crops. Biogas can be produced using anaerobic digesters. These plants can be fed with energy crops such as maize silage or biodegradable wastes including sewage sludge and food waste. During the process, an air-tight tank transforms biomass waste into methane producing renewable energy that can be used for heating, electricity, and many other operations that use any variation of an internal combustion engine, such as GE Jenbacher gas engines. There are two key processes: Mesophilic and Thermophilic digestion. In experimental work at University of Alaska Fairbanks, a 1000-litre digester using psychrophiles harvested from "mud from a frozen lake in Alaska" has produced 200–300 litres of methane per day, about 20–30 % of the output from digesters in warmer climates.

Landfill gas is produced by wet organic waste decomposing under anaerobic conditions in a landfill. The waste is covered and mechanically compressed by the weight of the material that is deposited from above. This material prevents oxygen exposure thus allowing anaerobic microbes to thrive. This gas builds up and is slowly released into the atmosphere if the landfill site has not been engineered to capture the gas. Landfill gas is hazardous for three key reasons. Landfill gas becomes explosive when it escapes from the landfill and mixes with oxygen. The lower explosive limit is 5% methane and the upper explosive limit is 15% methane. The methane contained within biogas is 20 times more potent as a greenhouse gas than is carbon dioxide. Therefore, uncontained landfill gas, which escapes into the atmosphere may significantly contribute to the effects of global warming. In addition, landfill gas impact in global warming, volatile organic compounds (VOCs) contained within landfill gas contribute to the formation of photochemical smog.

Composition

Typical composition of biogas[10]

Compound	Chem	%
Methane	CH_4	50–75
Carbon dioxide	CO_2	25–50
Nitrogen	N_2	0–10
Hydrogen	H_2	0–1
Hydrogen sulfide	H_2S	0–3
Oxygen	O_2	0–0

The composition of biogas varies depending upon the origin of the anaerobic digestion process. Landfill gas typically has methane concentrations around 50%. Advanced waste treatment technologies can produce biogas with 55–75% methane, which for reactors with free liquids can be increased to 80-90% methane using in-situ gas purification techniques As-produced, biogas also contains water vapor. The fractional volume of water vapor is a function of biogas temperature; correction of measured gas volume for both water vapor content and thermal expansion is easily done via a simple mathematic algorithm which yields the standardized volume of dry biogas.

In some cases, biogas contains siloxanes. These siloxanes are formed from the anaerobic decomposition of materials commonly found in soaps and detergents. During combustion of biogas containing siloxanes, silicon is released and can combine with free oxygen or various other elements in the combustion gas. Deposits are formed containing mostly silica (SiO_2) or silicates (Si_xO_y) and can also contain calcium, sulfur, zinc, phosphorus. Such white mineral deposits accumulate to a surface thickness of several millimeters and must be removed by chemical or mechanical means.

Practical and cost-effective technologies to remove siloxanes and other biogas contaminants are currently available.

Benefits

When biogas is used, many advantages arise. In North America, utilization of biogas would generate enough electricity to meet up to three percent of the continent's electricity expenditure. In addition, biogas could potentially help reduce global climate change. Normally, manure that is left to decompose releases two main gases that cause global climate change: nitrous dioxide and methane. Nitrous dioxide (NO_2) warms the atmosphere 310 times more than carbon dioxide and methane 21 times more than carbon dioxide. By converting cow manure into methane biogas via anaerobic digestion, the millions of cows in the United States would be able to produce one hundred billion kilowatt hours of electricity, enough to power millions of homes across the United States. In fact, one cow can produce enough manure in one day to generate three kilowatt hours of electricity; only 2.4 kilowatt hours of electricity are needed to power a single one hundred watt light bulb for one day. Furthermore, by converting cow manure into methane biogas instead of letting it decompose, we would be able to reduce global warming gases by ninety-nine million metric tons or four percent.

The 30 million rural households in China that have biogas digesters enjoy 12 benefits: saving fossil fuels, saving time collecting firewood, protecting forests, using crop residues for animal fodder instead of fuel, saving money, saving cooking time, improving hygienic conditions, producing high-quality fertilizer, enabling local mechanization and electricity production, improving the rural standard of living, and reducing air and water pollution.

Applications

Biogas can be utilized for electricity production on sewage works, in a CHP gas engine, where the waste heat from the engine is conveniently used for heating the digester; cooking; space heating; water heating; and process heating. If compressed, it can replace compressed natural gas for use in vehicles, where it can fuel an internal combustion engine or fuel cells and is a much more effective displacer of carbon dioxide than the normal use in on-site CHP plants.

Methane within biogas can be concentrated via a biogas upgrader to the same standards as fossil natural gas, which itself has had to go through a cleaning process, and becomes *biomethane.* If the local gas network allows for this, the producer of the biogas may utilize the local gas distribution networks. Gas must be very clean to reach pipeline quality, and must be of the correct composition for the local distribution network to accept. Carbon dioxide, water, hydrogen sulfide, and particulates must be removed if present.

Biogas Upgrading

Raw biogas produced from digestion is roughly 60% methane and 29% CO_2 with trace elements of H_2S, and is not high quality enough if the owner was planning on selling this gas or using it as fuel gas for machinery. The corrosive nature of H_2S alone is enough to destroy the internals of an expensive plant. The solution is the use of a biogas upgrading or purification process whereby contaminants in the raw biogas stream are absorbed or scrubbed, leaving 98% methane per unit volume of gas. There are four main methods of biogas upgrading, these include water washing, pressure swing absorption, selexol absorption, and amine gas treating. The most prevalent method is water washing where high pressure gas flows into a column where the carbon dioxide and other trace elements are scrubbed by cascading water running counter-flow to the gas. This arrangement could deliver 98% methane with manufacturers guaranteeing maximum 2% methane loss in the system. It takes roughly between 3-6% of the total energy output in gas to run a biogas upgrading system.

Biogas Gas-grid Injection

Gas-grid injection is the injection of biogas into the methane grid (natural gas grid). Injections includes biogas: until the breakthrough of micro combined heat and power two-thirds of all the energy produced by biogas power plants was lost (the heat), using the grid to transport the gas to customers, the electricity and the heat can be used for on-site generation resulting in a reduction of losses in the transportation of energy. Typical energy losses in natural gas transmission systems range from 1–2%. The current energy losses on a large electrical system range from 5–8%.

Biogas in Transport

If concentrated and compressed, it can also be used in vehicle transportation. Compressed biogas is becoming widely used in Sweden, Switzerland, and Germany. A biogas-powered train has been in service in Sweden since 2005. Biogas also powers automobiles and in 1974, a British documentary film entitled *Sweet as a Nut* detailed the biogas production process from pig manure, and how the biogas fueled a custom-adapted combustion engine. In 2007, an estimated 12,000 vehicles were being fueled with upgraded biogas worldwide, mostly in Europe.

Legislation

The European Union presently has some of the strictest legislation regarding waste management and landfill sites called the Landfill Directive. The United States legislates against landfill gas as it contains VOCs. The United States Clean Air Act and Title 40 of the Code of Federal Regulations (CFR) requires landfill owners to estimate the quantity of non-methane organic compounds (NMOCs) emitted. If the estimated NMOC emissions exceeds 50 tonnes per year, the landfill owner is required to collect the landfill gas and treat it to remove the entrained NMOCs. Treatment of the landfill gas is usually by combustion. Because of the remoteness of landfill sites, it is sometimes not economically feasible to produce electricity from the gas. However, countries such as the United Kingdom and Germany now have legislation in force that provides farmers with long-term revenue and energy security.

Development around the World

In the United States: With the many benefits of biogas, it is starting to become a popular source of energy and is starting to be utilized in the United States more. In 2003, the United States consumed

147 trillion BTU of energy from "landfill gas", about 0.6% of the total U.S. natural gas consumption. Methane biogas derived from cow manure is also being tested in the U.S. According to a 2008 study, collected by the *Science and Children* magazine, methane biogas from cow manure would be sufficient to produce 100 billion kilowatt hours enough to power millions of homes across America. Furthermore, methane biogas has been tested to prove that it can reduce 99 million metric tons of greenhouse gas emissions or about 4% of the greenhouse gases produced by the United States.

In Vermont, for example, biogas generated on dairy farms around the state is included in the CVPS Cow Power program. The Cow Power program is offered by Central Vermont Public Service Corporation as a voluntary tariff. Customers can elect to pay a premium on their electric bill, and that premium is passed directly to the farms in the program. In Sheldon, Vermont, Green Mountain Dairy has provided renewable energy as part of the Cow Power program. It all started when the brothers who own the farm, Bill and Brian Rowell, wanted to address some of the manure management challenges faced by dairy farms, including manure odor, and nutrient availability for the crops they need to grow to feed the animals. They installed an anaerobic digester to process the cow and milking center waste from their nine hundred and fifty cows to produce renewable energy, a bedding to replace sawdust, and a plant friendly fertilizer. The energy and environmental attributes are sold. On average, the system run by the Rowell brothers produces enough electricity to power three hundred to three hundred fifty other homes. The generator capacity is about three hundred kiloWatts.

In Hereford, Texas, cow manure is being used to power an ethanol power plant. By switching to methane biogas, the ethanol power plant has saved one thousand barrels of oil a day. Overall, the power plant has reduced transportation costs and will be opening many more jobs for future power plants that will be relying on biogas.

In the United Kingdom

There are currently around 60 non-sewage biogas plants in the UK, most are on-farm, but some larger facilities exist off-farm, which are taking food and consumer wastes. On the 5th October 2010, biogas was injected into the UK gas grid for the first time. Sewage from over 30,000 Oxfordshire homes is sent to Didcot sewage treatment works, where it is treated in an anaerobic digestor to produce biogas, which is then cleaned to provide gas for approximately 200 homes.

In Germany

Germany is Europe's biggest biogas producer as it is the market leader in biogas technology. In 2010 there were 5,905 biogas plants operating throughout the whole country, in which Lower Saxony, Bavaria and the eastern federal states are the main regions. Most of these plants are employed as power plants. Usually the biogas plants are directly connected with a CHP which produces electric power by burning the bio methane. The electrical power is then fed into the public power grid. In 2010 the total installed electrical capacity of these power plants was 2,291 MW. The electricity supply was approximately 12.8 TWh which is 12.6 per cent of the total generated renewable electricity. Biogas in Germany is primarily extracted by the co-fermentation of energy crops (In German science mostly the term 'NawaRo' is used that means 'nachwachsende Rohstoffe' = renewable resources) mixed with excrements or rather manure, the main crop utilized is corn. Furthermore bio waste and industrial and agricultural residues such as waste from the food industry are expended. Therefore biogas production in Germany differs highly from the methods of production in the UK with regards to the selection of the raw materials that biogas is extracted from, because mainly landfill gas is used in the UK's biogas plants.

Biogas production in Germany has developed rapidly over the last 20 years. The main reason for this development is the legally created frameworks. Governmental support of renewable energies started at the beginning of the 1990s with the Law on Electricity Feed (StrEG). This law guaranteed the producers of energy from renewable sources the feed into the public power grid, thus the power companies were forced to take all produced energy from independent private producers of green energy. In 2002 the Law on Electricity Feed was replaced by the Renewable Energy Source Act (EEG). This law even guaranteed a fixed compensation for the produced electric power over 20 years. The amount of ca. 0.08 Euro gave in particular framers the opportunity to become an energy supplier and gaining a further source of income in the same place. The German agricultural biogas production was given a further push in 2004 by implementing the so-called NawaRo-Bonus. This is a special bonus payment given for the usage of renewable resources i.e. energy crops. In 2007 the German government stressed its intention to invest further effort and support in improving the renewable energy supply to provide an answer on growing climate challenges and increasing oil prices by the 'Integrated Climate and Energy Programme'.

This continual trend of renewable energy promotion induces a number of challenges facing the management and organisation of renewable energy supply that has also several impacts on the biogas production. The first challenge to be noticed is the high area-consuming of the biogas electric power supply. In 2011 energy crops for biogas production consumed an area of circa 800,000 ha in Germany. This high demand of agricultural areas generates new competitions with the food industries that did not exist yet. Moreover new industries and markets were created in predominately rural regions entailing different new players with an economic, political and civil background. Their influence and acting has to be governed to gain all advantages this new source of energy is offering. Finally biogas will furthermore play an important role in the German renewable energy supply if good governance is focused.

In the Indian Subcontinent

In Pakistan, India, and bangladesh biogas produced from the anaerobic digestion of manure in small-scale digestion facilities is called gobar gas; it is estimated that such facilities exist in over two million households in India and in hundreds of thousands in Pakistan, particularly North Punjab, due to the thriving population of livestock. It has become popular source of fuel in many parts of Nepal. The digester is an airtight circular pit made of concrete with a pipe connection. The manure is directed to the pit, usually directly from the cattle shed. The pit is then filled with a required quantity of wastewater. The gas pipe is connected to the kitchen fireplace through control valves. The combustion of this biogas has very little odour or smoke. Owing to simplicity in implementation and use of cheap raw materials in villages, it is one of the most environmentally sound energy sources for rural needs. One type of these system is the Sintex Digester. Some designs use vermiculture to further enhance the slurry produced by the biogas plant for use as compost. In order to create awareness and associate the people interested in biogas, an association "Indian Biogas Association" is formed. The "Indian Biogas Association" aspires to be a unique blend of; nationwide operators, manufacturers and planners of biogas plants, and representatives from science and research. The association was founded in 2010 and is now ready to start mushrooming. The sole motto of the association is "propagating Biogas in a sustainable way".

The Deenabandhu Model is a new biogas-production model popular in India. (*Deenabandhu* means "friend of the helpless.") The unit usually has a capacity of 2 to 3 cubic metres. It is constructed using bricks or by a ferrocement mixture. In India, the brick model costs slightly more than the ferrocment model; however, India's Ministry of New and Renewable Energy offers some subsidy per model constructed.

In Pakistan, PAK-Energy Solution from University of Engineering and Technology, Lahore has taken the most innovative and responsible initiatives in biogas technology. In this regard, the company is also awarded by 1st prize in "Young Entrepreneur Business Plan Competition" jointly organized by Punjab Govt. & LCCI and "Battle of Business Giants" in Techno'Fest 11. The company is aiming to install 70,000 biogas plants in next 3 years. They have designed and developed Uetians Hybrid Model, in which they have combined fixed dome and floating drums and Uetians Triplex Model. Both of these designs have been innovated first time in the world. 10 years of experience of an expert team of young engineers & entrepreneurs has put together in research & development along with senior Ph.D advisory board. Moreover, Pakistan Dairy Development Company has also taken an initiative to develop this kind of alternative source of energy for Pakistani farmers. Biogas is now running diesel engines, gas generators, kitchen ovens, geysers, and other utilities in Pakistan. In Nepal, the government provides subsidies to build biogas plant.

China

The Chinese have been experimenting with the applications of biogas since 1958. Around 1970, China had installed 6,000,000 digesters in an effort to make agriculture more efficient. During the last years the technology has met high growth rates.

In Developing Nations

Domestic biogas plants convert livestock manure and night soil into biogas and slurry, the fermented manure. This technology is feasible for small holders with livestock producing 50 kg manure per day, an equivalent of about 6 pigs or 3 cows. This manure has to be collectable to mix it with water and feed it into the plant. Toilets can be connected. Another precondition is the temperature that affects the fermentation process. With an optimum at 36 C° the technology especially applies for those living in a (sub) tropical climate. This makes the technology for small holders in developing countries often suitable.

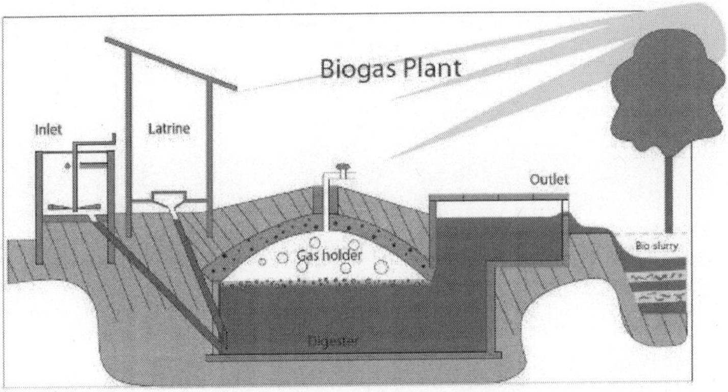

Figure: *Simple sketch of household biogas plant*

Depending on size and location, a typical brick made fixed dome biogas plant can be installed at the yard of a rural household with the investment between 300 to 500 US $ in Asian countries and up to 1400 US $ in the African context. A high quality biogas plant needs minimum maintenance costs and can produce gas for at least 15–20 years without major problems and re-investments. For the user, biogas provides clean cooking energy, reduces indoor air pollution, and reduces the time needed for traditional biomass collection, especially for women and children. The slurry is a clean organic fertilizer that potentially increases agricultural productivity.

Domestic biogas technology is a proven and established technology in many parts of the world, especially Asia. Several countries in this region have embarked on large-scale programmes on domestic biogas, such as China and India. The Netherlands Development Organisation, SNV, supports national programmes on domestic biogas that aim to establish commercial-viable domestic biogas sectors in which local companies market, install and service biogas plants for households. In Asia, SNV is working in Nepal, Vietnam, Bangladesh, Bhutan, Cambodia, Lao PDR, Pakistan and Indonesia, and in Africa; Rwanda, Senegal, Burkina Faso, Ethiopia, Tanzania, Uganda, Kenya, Benin and Cameroon.

10

Ecological Sanitation

Ecological sanitation, also known as ecosan or eco-san, are terms coined to describe a form of sanitation that usually involves urine diversion and the recycling of water and nutrients contained within human wastes back into the local environment. An ecological sanitation (ecosan) viewpoint sees human waste and wastewater as an opportunity. When properly designed and operated, ecosan systems provide a hygienically safe, economical, and closed-loop system to convert human wastes into nutrients to be returned to the soil, and water to be returned to the land. Alternatively, solid wastes are converted into a biofuel. The primary application for ecosan systems has been in rural areas where connection to a sanitary sewer system is not possible, or where water supplies are very limited.

The main objectives of ecological sanitation are:

- To reduce the health risks related to sanitation, contaminated water and waste
- To prevent the pollution of surface and ground water
- To reuse nutrients or energy contained within wastes.

Reuse-oriented Sanitation Approaches

The recovery and use of urine and feces has been practiced by almost all cultures. The reuse was not limited to agricultural production. The Romans, for example, were aware of the bleaching attribute of the ammonia within urine and used it to whiten clothing.

The most widely known reuse in agriculture has occurred in China. Reportedly, the Chinese were aware of the benefits of using excreta in crop production before 500 B.C., enabling them to sustain more people at a higher density than any other system of agriculture. The value of "night soil" as a fertilizer was recognized with well-developed systems in place to enable the collection of excreta from cities and its transportation to fields. However, its use promoted disease to such an extent that in Chinese cuisine almost all vegetables are thoroughly cooked.

Elaborate systems were developed in urban centers of Yemen enabling the separation of urine and excreta even in multi-story buildings. Feces were collected from toilets via vertical drop shafts, while urine did not enter the shaft but passed instead along a channel leading through the wall to the outside where it evaporated. Here, feces were not used in agriculture but were dried and burnt as a biofuel. In Mexico and Peru, both the Aztec and Inca cultures collected human excreta for agricultural use. In Peru, the Incas had a high regard for excreta as a fertilizer, which was stored, dried and pulverized to be utilized when planting maize. In the Middle Ages, the use of excreta and greywater was the norm. European cities were rapidly urbanizing and sanitation was becoming an increasingly serious problem, whilst at the same time the cities themselves were becoming an increasingly important source of agricultural nutrients. The practice of using the nutrients in excreta and wastewater for agriculture therefore continued in Europe into the middle of the 19th Century. Farmers, recognizing the value of excreta, were eager to get these fertilizers to increase production and urban sanitation benefited.

The increasing number of research and demonstration projects for excreta reuse carried out in Sweden from the 1980s to the early 21st century aimed at developing hygienically safe closed loop sanitation systems. Similar lines of research began elsewhere, for example in Zimbabwe, in the Netherlands, Norway and Germany. These closed-loop sanitation systems became popular under the name "ecosan", "dewats", "desar", and other abbreviations. They placed their emphasis on the hygenisation of the contaminated flow streams, and shifted the concept from waste disposal to resource conservation and safe reuse.

Technologies of Ecosan Systems

Determining ecosan systems as ecological sanitation is not easy, for it is not just one specific technology, but a new approach based on an ecosystem-oriented view of material flows. The following diagram gives an overview of the different collection, treatment and reuse possibilities of the five flow streams considered in ecological sanitation systems:

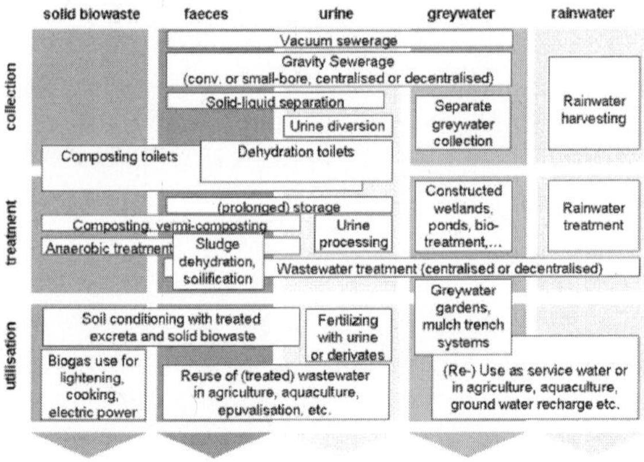

***Figure:** Further information on ecosan technologies can be found in "Ecological Sanitation" by Winblad et al., in "Toilets that make compost" by Peter Morgan or in the gtz-ecosan technical data sheets, among other relevant literature.*

Environmental Design

Environmental design is the process of addressing surrounding environmental parameters when devising plans, programs, policies, buildings, or products. Classical prudent design may have always considered environmental factors; however, the environmental movement beginning in the 1940s has made the concept more explicit.

Environmental design can also refer to the applied arts and sciences dealing with creating the human-designed environment. These fields include architecture, geography, u rban planning, landscape architecture, and interior design. Environmental design can also encompass interdisciplinary areas such as historical preservation and lighting design. In terms of a larger scope, environmental design has implications for the industrial design of products: innovative automobiles, wind-electricity generators, solar-electric equipment, and other kinds of equipment could serve as examples. Currently, the term has expanded to apply to ecological and sustainability issues.

History

The first traceable concepts of environmental designs focused primarily on solar heating, which began in Ancient Greece around 500 BCE. At the time, most of Greece had exhausted its supply of wood for fuel, leading architects to design houses that would capture the solar energy of the sun. The Greeks understood that the position of the sun varies throughout the year. For a latitude of 40 degrees in summer the sun is high in the south, at an angle of 70 degrees at the zenith, while in winter, the sun travels a lower trajectory, with a zenith of 26 degrees. Greek houses were built with south-facing faηades which received little to no sun in the summer but would receive full sun in the winter, warming the house. Additionally, the southern orientation also protected the house from the colder northern winds. This clever arrangement of buildings influenced the use of the grid pattern of ancient cities. With the North-South orientation of the houses, the streets of Greek cities mainly ran East-West.

The practice of solar architecture continued with the Romans, who similarly had deforested much of their native Italian Peninsula by the first century BCE. The Roman *heliocaminus*, literally 'solar furnace', functioned with the same aspects of the earlier Greek houses. The numerous public baths were oriented to the south. Roman architects added glass to windows to allow for the passage of light and to conserve interior heat as it could not escape. The Romans also used greenhouses to grow crops all year long and to cultivate the exotic plants coming from the far corners of the Empire. Pliny the Elder wrote of greenhouses that supplied the kitchen of the Emperor Tiberius during the year.

Along with the solar orientation of buildings and the use of glass as a solar heat collector, the ancients knew other ways of harnessing solar energy. The Greeks, Romans and Chinese developed curved mirrors that could concentrate the sun's rays on an object with enough intensity to make it burn in seconds. The solar reflectors were often made of silver, copper or polished brass. Early roots of modern environmental design began in the late 19th Century with writer/ designer William Morris, who rejected the use of industrialized materials and processes in wallpaper, fabrics and books his studio produced. He and others, such as John Ruskin felt that the industrial revolution would lead to harm done to nature and workers.

The narrative of Phil Cousineau's documentary film *Ecological Design: Inventing the Future* asserts that in the decades after World War II, "The world was forced to confront the dark shadow of science and industry." From the middle of the twentieth century, thinkers like Buckminster Fuller have acted as catalysts for a broadening and deepening of the concerns of environmental designers. Nowadays, energy efficiency, appropriate technology, organic horticulture and agriculture, land restoration, New Urbanism, and ecologically sustainable energy and waste systems are recognized considerations or options and may each find application.

By integrating renewable energy sources such as solar photovoltaic, solar thermal, and even geothermal energy into structures, it is possible to create zero emission buildings, where energy consumption is self-generating and non-polluting. It is also possible to construct "energy-plus buildings" which generate more energy than they consume, and the excess could then be sold to the grid. In the United States, the LEED Green Building Rating System rates structures on their environmental sustainability.

Examples

Examples of the environmental design process include use of roadway noise computer models in design of noise barriers and use of roadway air dispersion models in analyzing and designing urban highways. Designers consciously working within this more recent framework of philosophy and practice seek a blending of nature and technology, regarding ecology as the basis for design. Some believe that strategies of conservation, stewardship, and regeneration can be applied at all levels of scale from the individual building to the community, with benefit to the human individual and local and planetary ecosystems.

Specific examples of large scale environmental design projects include:

- Boston Transportation Planning Review
- BART - Bay Area Rapid Transit System Daly City Turnback project and airport extension.
- Metropolitan Portland, Oregon light rail system

Environmental Engineering Law

Environmental engineering law is a profession that requires an expertise in both environmental engineering and law. This field includes professionals with both a legal and environmental engineering education. This dual educational requirement is typically satisfied through an ABET accredited degree in environmental engineering and an ABA accredited law degree. Likewise, this profession requires both licensure in professional environmental engineering and admittance to one bar. Environmental engineering law is the professional application of law, science and engineering principles to improve the environment (air, water, and/or land resources), to provide healthy water, air, and land for human habitation and for other organisms, and to remediate polluted sites. Environmental engineering lawyers seek to promote the advancement of technical engineering knowledge in the legal profession and to enhance informed legal analysis of complex environmental matters.

Practice Areas

Environmental engineering law professionals offer a sound knowledge base in the fields of both environmental engineering and law to address complex environmental problems which demand both professional technical practice and legal expertise. Areas of practice are continually expanding, but frequently include complex land transactions, such as:

- Brownfields redevelopment
- Asbestos baseline survey and building revaluation due to forthcoming asbestos abatements
- Soil contamination assessment & remediation, the development of a remedial action workplan (RAWP) and engineering controls, including an environmental land use restriction (ELUR)
- Total maximum daily load (TMDL) nutrient loading studies (ex. for NPDES wastewater discharges) and regulatory negotiation of nutrients discharge limits from waste treatment plants, such as phosphorus and nitrogen.

Environmental Impact Assessment

An environmental impact assessment is an assessment of the possible positive or negative impact that a proposed project may have on the environment, together consisting of the natural, social and economic aspects. The purpose of the assessment is to ensure that decision makers consider the ensuing environmental impacts when deciding whether to proceed with a project. The International Association for Impact Assessment (IAIA) defines an environmental impact assessment as "the process of identifying, predicting, evaluating and mitigating the biophysical, social, and other relevant effects of development proposals prior to major decisions being taken and commitments made." EIAs are unique in that they do not require adherence to a predetermined environmental outcome, but rather they require decision-makers to account for environmental values in their decisions and to justify those decisions in light of detailed environmental studies and public comments on the potential environmental impacts of the proposal.

EIAs began to be used in the 1960s as part of a rational decision making process. It involved a technical evaluation that would lead to objective decision making. EIA was made legislation in the US in the National Environmental Policy Act (NEPA) 1969. It has since evolved as it has been used increasingly in many countries around the world. As per Jay *et al.*(2006), EIA as it is practiced today, is being used as a decision aiding tool rather than decision making tool. There is growing dissent on the use of EIA as its influence on development decisions is limited and there is a view it is falling short of its full potential.There is a need for stronger foundation of EIA practice through training for practitioners, guidance on EIA practice and continuing research.

EIAs have often been criticized for having too narrow spatial and temporal scope. At present no procedure has been specified for determining a system boundary for the assessment. The system boundary refers to 'the spatial and temporal boundary of the proposal's effects'. This boundary is determined by the applicant and the lead assessor, but in practice, almost all EIAs address the direct, on-site effects alone.

However, as well as direct effects, developments cause a multitude of indirect effects through consumption of goods and services, production of building materials and machinery, additional land use for activities of various manufacturing and industrial services, mining of resources etc. The indirect effects of developments are often an order of magnitude higher than the direct effects assessed by EIA. Large proposals such as airports or ship yards cause wide ranging national as well as international environmental effects, which should be taken into consideration during the decision-making process.

Broadening the scope of EIA can also benefit threatened species conservation. Instead of concentrating on the direct effects of a proposed project on its local environment some EIAs used a landscape approach which focused on much broader relationships between the entire population of a species in question. As a result, an alternative that would cause least amount of negative effects to the population of that species as a whole, rather than the local subpopulation, can be identified and recommended by EIA.

There are various methods available to carry out EIAs, some are industry specific and some general methods:

- Industrial products - Product environmental life cycle analysis (LCA) is used for identifying and measuring the impact on the environment of industrial products. These EIAs consider technological activities used for various stages of the product:

extraction of raw material for the product and for ancillary materials and equipment, through the production and use of the product, right up to the disposal of the product, the ancillary equipment and material.

- Genetically modified plants - There are specific methods available to perform EIAs of genetically modified plants. Some of the methods are GMP-RAM, INOVA etc.
- Fuzzy Arithmetic - EIA methods need specific parameters and variables to be measured to estimate values of impact indicators. However many of the environment impact properties cannot be measured on a scale e.g. landscape quality, lifestyle quality, social acceptance etc. and moreover these indicators are very subjective. Thus to assess the impacts we may need to take the help of information from similar EIAs, expert criteria, sensitivity of affected population etc. To treat this information, which is generally inaccurate, systematically, fuzzy arithmetic and approximate reasoning methods can be utilised. This is called as a fuzzy logic approach.

At the end of the project, an EIA should be followed by an audit. An EIA audit evaluates the performance of an EIA by comparing actual impacts to those that were predicted. The main objective of these audits is to make future EIAs more valid and effective. The two main considerations are:

- scientific - to check the accuracy of predictions and explain errors.
- management- to assess the success of mitigation in reducing impacts.

Some people believe that audits be performed as a rigorous scientific testing of the null hypotheses. While some believe in a simpler approach where you compare what actually occurred against the predictions in the EIA document.

After an EIA, the precautionary and polluter pays principles may be applied to prevent, limit, or require strict liability or insurance coverage to a project, based on its likely harms. Environmental impact assessments are sometimes controversial.

Application

Australia: The history of EIA in Australia could be linked to the enactment of the U.S. National Environment Policy Act (NEPA) in 1970, which made the preparation of environmental impact statements a requirement. In Australia, one might say that the EIA procedures were introduced at a State Level prior to that of the Commonwealth (Federal), with a majority of the states having divergent views to the Commonwealth. One of the pioneering states was New South Wales, whose State Pollution Control Commission issued EIA guidelines in 1974. At a Commonwealth (Federal) level, this was followed by passing of the Environment Protection (Impact of Proposals) Act in 1974. The Environment Protection and Biodiversity Conservation Act 1999 (EPBC) superseded the Environment Protection (Impact of Proposals) Act 1974 and is the current central piece for EIA in Australia on a Commonwealth (Federal) level. An important point to note is that this Commonwealth Act does not affect the validity of the States and Territories environmental and development assessments and approvals; rather the EPBC runs as a parallel to the State/Territory Systems. Overlap between federal and state requirements is addressed via bilateral agreements or one off accredition of state processes, as provided for in the EPBC Act.

The Commonwealth Level

The EPBC Act provides a legal framework to protect and manage nationally and internationally important flora, fauna, ecological communities and heritage places-defined in the EPBC Act as matters of 'national environmental significance'. Following are the eight matters of 'national environmental significance' to which the EPBC ACT applies:

- World Heritage sites;
- National Heritage places;
- RAMSAR wetlands of international significance;
- Listed threatened species and ecological communities;
- Migratory species protected under international agreements;
- The Commonwealth marine environment;
- Nuclear actions (including uranium mining); and
- National Heritage.

In addition to this, the EPBC Act aims at providing a streamlined national assessment and approval process for activities. These activities could be by the Commonwealth, or its agents, anywhere in the world or activities on Commonwealth land; and activities that are listed as having a 'significant impact' on matters of 'national environment significance'.

The EPBC Act comes into play when a person (a 'proponent') wants an action (often called a 'proposal' or 'project') assessed for environmental impacts under the EPBC Act, he or she must refer the project to the Department of Environment, Water, Heritage and the Arts (Australia). This 'referral' is then released to the public, as well as relevant state, territory and Commonwealth ministers, for comment on whether the project is likely to have a significant impact on matters of national environmental significance. The Department of Environment, Water, Heritage and the Arts assess the process and makes recommendation to the minister or the delegate for the feasibility. The final discretion on the decision remains of the minister, which is not solely based on matters of 'national environmental significance' but also the consideration of social and economic impact of the project.

The Australian Government environment minister cannot intervene in a proposal if it has no significant impact on one of the eight matters of 'national environmental significance' despite the fact that there may be other undesirable environmental impacts. This is primarily due to the division of powers between the States and the Federal government and due to which the Australian Government environment minister cannot overturn a state decision.

There are strict civil and criminal penalties for the breach of EPBC Act. Depending on the kind of breach, civil penalty (maximum) may go up to $550,000 for an individual and $5.5 million for a body corporate, or for criminal penalty (maximum) of seven years imprisonment and/or penalty of $46,200.

The State and Territory Level

Australian Capital Territory (ACT): EIA in ACT is administered with the help of Part 4 of the Land (Planning and Environment) Act 1991 (Land Act) and Territory Plan (plan for land-use).

New South Wales (NSW)

In New South Wales, the Environment Planning Assessment Act 1979 (EPA) establishes three pathways for EIA. The first is under Part 3A of the Act where Major Projects require an environmental assessment. The second is under Part 4 of the Act dealing with development control. If a project does not require approval under Part 3A or Part 4 it is then potentially captured by the third pathway, Part 5 dealing with environment impact assessment.

Northern Territory (NT)

The EIA process in Northern Territory is chiefly administered under the Environmental Assessment Act (EEA). Although EEA is the primary tool for EIA in Northern Territory, there are further provisions for proposals in the Inquiries Act 1985 (NT).

Queensland (QLD)

There are four main EIA processes in Queensland. Firstly, under the Integrated Planning Act 1997 (IPA) for development projects other than mining. Secondly, under the Environmental Protection Act 1994 (EP Act) for some mining and petroleum activities. Thirdly, under the State Development and Public Works Organization Act 1971 (State Development Act) for 'significant projects'. Finally, Environment Protection and Biodiversity Conservation Act 1999 (Cth) for 'controlled actions'.

South Australia (SA)

The local governing tool for EIA in South Australia is the Development Act 1993. There are three levels of assessment possible under the Act in the form of an environment impact statement (EIS), a public environmental report (PER) or a Development Report (DR).

Tasmania (TAS)

In Tasmania, an integrated system of legislation is used to govern development and approval process, this system is a mixture of the Environmental Management and Pollution Control Act 1994 (EMPCA), Land Use Planning and Approvals Act 1993 (LUPAA), State Policies and Projects Act 1993 (SPPA), and Resource Management and Planning Appeals Tribunal Act 1993.

Victoria (VIC)

The EIA process in Victoria is intertwined with the Environment Effects Act 1978 and the Ministerial Guidelines for Assessment of Environmental Effects (made under the s. 10 of the EE Act).

Western Australia (WA)

The Environmental Protection Act 1986 (Part 4) provides the legislative framework for the EIA process in Western Australia. The EPA Act oversees the planning and development proposals and assesses their likely impacts on the environment.

Canada

This section does not cite any references or sources. Please help improve this section by adding citations to reliable sources. Unsourced material may be challenged and removed. *(November 2010)*

The *Canadian Environmental Assessment Act* (CEAA) is the legal basis for the federal environmental assessment (EA) process. CEAA came into force in 1995. Legislative amendments were introduced in 2001 and came into force on October 30, 2003. EA is defined as a planning tool to identify, understand, assess and mitigate, where possible, the environmental effects of a project. Under the CEAA, all federal government departments and agencies are required to undertake an EA for projects relating to a physical work and for any proposed physical activity listed in the Inclusion List Regulations where it exercises one or more of the following CEAA triggers:

- Proposes or undertakes a project
- Grants money or any other form of financial assistance to a project
- Grants an interest in the land to enable a project to be carried out
- Exercises a regulatory duty in relation to a project, such as issuing a permit or license that is included in the Law List Regulations.

If a federal government department or agency exercises one or more of the above-mentioned triggers, it becomes a Responsible Authority (RA) under CEAA. As an RA, the federal department or agency in question must ensure that an EA is carried out in accordance with the CEAA and must consider the EA findings before a decision is made that could allow the project to proceed.

China

The Environmental Impact Assessment Law (EIA Law) requires an environmental impact assessment to be completed prior to project construction. However, if a developer completely ignores this requirement and builds a project without submitting an environmental impact statement, the only penalty is that the environmental protection bureau (EPB) may require the developer to do a make-up environmental assessment. If the developer does not complete this make-up assessment within the designated time, only then is the EPB authorized to fine the developer. Even so, the possible fine is capped at a maximum of about US$25,000, a fraction of the overall cost of most major projects. The lack of more stringent enforcement mechanisms has resulted in a significant percentage of projects not completing legally required environmental impact assessments prior to construction.

China's State Environmental Protection Administration (SEPA) used the legislation to halt 30 projects in 2004, including three hydropower plants under the Three Gorges Project Company. Although one month later (Note as a point of reference, that the typical EIA for a major project in the USA takes one to two years.), most of the 30 halted projects resumed their construction, reportedly having passed the environmental assessment, the fact that these key projects' construction was ever suspended was notable.

A joint investigation by SEPA and the Ministry of Land and Resources in 2004 showed that 30-40% of the mining construction projects went through the procedure of environment impact assessment as required, while in some areas only 6-7% did so. This partly explains why China has witnessed so many mining accidents in recent years.

SEPA alone cannot guarantee the full enforcement of environmental laws and regulations, observed Professor Wang Canfa, director of the centre to help environmental victims at China University of Political Science and Law. In fact, according to Wang, the rate of China's environmental laws and regulations that are actually enforced is estimated to be barely 10%.

Egypt

Environmental Impact Assessment (EIA) EIA is implemented in Egypt under the umbrella of the Ministry of state for environmental affairs. The Egyptian Environmental Affairs Agency (EEAA) is responsible for the EIA services.

In June 1997, the responsibility of Egypt's first full time Minister of State for Environmental Affairs was assigned as stated in the Presidential Decree no.275/1997. From thereon, the new ministry has focused, in close collaboration with the national and international development partners, on defining environmental policies, setting priorities and implementing initiatives within a context of sustainable development.

According to the Law 4/1994 for the Protection of the Environment, the Egyptian Environmental Affairs Agency (EEAA) was restructured with the new mandate to substitute the institution initially established in 1982. At the central level, EEAA represents the executive arm of the Ministry. The purpose of EIA is to ensure the protection and conservation of the environment and natural resources including human health aspects against uncontrolled development. The long-term objective is to ensure a sustainable economic development that meets present needs without compromising future generations ability to meet their own needs. EIA is an important tool in the integrated environmental management approach. EIA must be performed for new establishments or projects and for expansions or renovations of existing establishments according to the Law for the Environment.

EU

The European Union has established a mix of mandatory and discretionary procedures to assess environmental impacts. European Union Directive (85/337/EEC) on Environmental Impact Assessments (known as the *EIA Directive*) was first introduced in 1985 and was amended in 1997. The directive was amended again in 2003, following EU signature of the 1998 Aarhus Convention. In 2001, the issue was enlarged to the assessment of plans and programmes by the so called *Strategic Environmental Assessment (SEA) Directive* (2001/42/EC), which is now in force. Under the EU directive, an EIA must provide certain information to comply. There are seven key areas that are required:

1. Description of the project
    - Description of actual project and site description
    - Break the project down into its key components, i.e. construction, operations, decommissioning
    - For each component list all of the sources of environmental disturbance
    - For each component all the inputs and outputs must be listed, e.g., air pollution, noise, hydrology
2. Alternatives that have been considered
    - Examine alternatives that have been considered
    - Example: in a biomass power station, will the fuel be sourced locally or nationally?
3. Description of the environment
    - List of all aspects of the environment that may be affected by the development
    - Example: populations, fauna, flora, air, soil, water, humans, landscape, cultural heritage
    - This section is best carried out with the help of local experts, e.g. the RSPB in the UK
4. Description of the significant effects on the environment
    - The word significant is crucial here as the definition can vary

- o 'Significant' needs to be defined
- o The most frequent method used here is use of the Leopold matrix
- o The matrix is a tool used in the systematic examination of potential interactions
- o Example: in a windfarm development a significant impact may be collisions with birds

5. Mitigation
   - o This is where EIA is most useful
   - o Once section 4 has been completed it will be obvious where the impacts will be greatest
   - o Using this information ways to avoid negative impacts should be developed
   - o Best working with the developer with this section as they know the project best
   - o Using the windfarm example again construction could be out of bird nesting seasons
6. Non-technical summary (EIS)
   - o The EIA will be in the public domain and be used in the decision making process
   - o It is important that the information is available to the public
   - o This section is a summary that does not include jargon or complicated diagrams
   - o It should be understood by the informed lay-person
7. Lack of know-how/technical difficulties
   - o This section is to advise any areas of weakness in knowledge
   - o It can be used to focus areas of future research
   - o Some developers see the EIA as a starting block for poor environmental management

The Netherlands

EIA was implemented in Dutch legislation on September 1, 1987. The categories of projects that require an EIA are summarised in Dutch legislation, the Wet milieubeheer. The use of thresholds for activities makes sure that EIA is obligatory for those activities that may have considerable impacts on the environment.

For projects and plans that fit these criteria, a EIA report is required. The EIA report defines a.o. the proposed initiative, it makes clear the impact of that initiative on the environment and compares this with the impact of possible alternatives with less a negative impact.

Hong Kong

EIA in Hong Kong, since 1998, is regulated by the *Environmental Impact Assessment Ordinance 1997.* The original proposal to construct the Lok Ma Chau Spur Line overground across the Long Valley failed to get through EIA, and the KCR Corporation had to change its plan and build the railway underground. In April 2011, the EIA of the Hong Kong section of the Hong Kong-Zhuhai-Macau Bridge was found to have breached the ordinance, and was declared unlawful. The appeal by the government was allowed in September 2011. However, it was estimated that this EIA court case had increased the construction cost of the Hong Kong section of the bridge by HK$6.5 billion in money-of-the-day prices.

India

The Ministry of Environment and Forests (MoEF) of India have been in a great effort in Environmental Impact Assessment in India. The main laws in nation are Water Act(1974), The Indian Wildlife (Protection) Act (1972), The Air (Prevention and Control of Pollution) Act (1981) and The Environment (Protection) Act (1986). The responsible body for this is Central Pollution Control Board. Environmental Impact Assessment (EIA) studies need a significant amount of primary and secondary environmental data. The primary data are those which need to be collected in the field to define the status of environment (like air quality data, water quality data etc.). The secondary data are those data which have been collected over the years and can be used to understand the existing environmental scenario of the study area. The environmental impact assessment (EIA) studies are conducted over a short period of time and therefore the understanding the environmental trends based on few months of primary data has its own limitations. Ideally, the primary data has to be considered along with the secondary data for complete understanding of the existing environmental status of the area. In many EIA studies, the secondary data needs could be as high as 80% of the total data requirement. EIC is the repository of one stop secondary data source for environmental impact assessment in India.

The Environmental Impact Assessment (EIA) experience in India indicates that the lack of timely availability of reliable and authentic environmental data has been a major bottle neck in achieving the full benefits of EIA. The environment being a multi-disciplinary subject, a multitude of agencies is involved in collection of environmental data. However, there is no single organization in India which tracks the data available amongst these agencies and makes it available in one place, in a form and manner required by practitioners in the field of environmental impact assessment in India. Further, the environmental data is not available in value added forms that can enhance the quality of the EIA. This in turn adversely affects the time and efforts required for conducting the environmental impact assessments (EIAs) by project proponents and also timely environmental clearances by the regulators. With this background, Environmental Information Centre (EIC) has been set up to serve as a professionally managed clearing house of environmental information that can be used by MoEF, project proponents, consultants, NGOs and other stakeholders involved in the process of environmental impact assessment in India. EIC caters to the need of creating and disseminating of organized environmental data for various developmental initiatives all over the country. EIC stores data in GIS format and makes it available to all environmental impact assessment studies and to EIA stakeholders in a cost effective and timely manner.

Malaysia

In Malaysia, Section 34A, Environmental Quality Act, 1974 requires developments that have significant impact to the environment are required to conduct the Environmental impact assessment.

Nepal

In Nepal, EIA has been integrated in major development projects since the early 1980s. In the planning history ofNepal, the sixth plan (1980–85), for the first time, recognized the need for EIA with the establishment of Environmental Impact Study Project (EISP) under the Department of Soil Conservation in 1982 to develop necessary instruments for integration of EIA in infrastructure development projects. However, the government of Nepal enunciated environment conservation related policies in the seventh plan (NPC, 1985–1990). In order to enforce this policy and make necessary arrangements, a series of guidelines were developed, thereby incorporating the elements of environmental factors right from the project formulation stage of the development plans and projects and to avoid or minimize adverse effects on the ecological system. In addition, it has also emphasized that EIAs of industry, tourism, water resources, transportation, urbanization, agriculture, forest and other developmental projects be conducted.

In Nepal, the government's Environmental Impact Assessment Guideline of 1993 inspired the enactment of the Environment Protection Act (EPA) of 1997 and the Environment Protection Rules (EPR) of 1997(EPA and EPR have been enforced since 24 and 26 June 1997 respectively in Nepal) to internalizing the environmental assessment system. The process institutionalized the EIA process in development proposals and enactment, which makes the integration of IEE and EIA legally binding to the prescribed projects. The projects, requiring EIA or IEE, are included in Schedules 1 and 2 of the EPR, 1997 (GoN/MoLJPA 1997). Progresses were made in the Environmental protection issue during the 8th five year plan (1992–1997). The following development in Environmental protection were achieved during that time:

- Formulation of Environmental Protection Act 1997
- Establishment of Ministry of Environment
- Development of National Environmental Policies and Action Plan, EIA guidelines developed
- Consideration of environmental concerns in hydropower projects
- Development of industrial, irrigation and agricultural policies that undertook environmental concerns

New Zealand

In New Zealand, EIA is usually referred to as *Assessment of Environmental Effects* (AEE). The first use of EIA's dates back to a Cabinet minute passed in 1974 called Environmental Protection and Enhancement Procedures. This had no legal force and only related to the activities of government departments. When the Resource Management Act was passed in 1991, an EIA was required as part of a resource consent application. Section 88 of the Act specifies that the AEE must include "such detail as corresponds with the scale and significance of the effects that the activity may have on the environment". While there is no duty to consult any person when making a resource consent application (Sections 36A and Schedule 4), proof of consultation is almost certain to be required by local councils when making a decision about whether or not to publicly notifiy the consent application under Section 93.

Russian Federation

Russia holds the world's largest natural gas reserves, the second largest coal reserves, and the eighth largest oil reserves. Russia is also the world's largest exporter of natural gas, the second largest oil exporter and the third largest energy consumer.

As of 2004, the state authority responsible for conducting the State EIA in Russia has been split between two Federal bodies: 1) Federal service for monitoring the use of natural resources – a part of the Russian Ministry for Natural Resources and Environment and 2) Federal Service for Ecological, Technological and Nuclear Control . The two main pieces of environmental legislation in Russia are: The Federal Law 'On Ecological Expertise, 1995 and the 'Regulations on Assessment of Impact from Intended Business and Other Activity on Environment in the Russian Federation, 2000.

Federal Service for Monitoring the Use of Natural Resources

In 2006, the parliament committee on ecology in conjunction with the Ministry for Natural Resources and Environment, created a working group to prepare a number of amendments to existing legislation in order to cover such topics as stringent project documentation for building of potentially environmentally damaging objects as well as building of projects on the territory of protected areas. There has been some success in this area, as evidenced from abandonment of plans to construct a gas pipe-line through the only remaining habitat of the critically endangered Amur leopard in the Russian Far East.

Federal Service for Ecological, Technological and Nuclear Control

The government's decision to hand over control over several important procedures, including state EIA in the field of all types of energy projects, to the Federal Service for Ecological, Technological and Nuclear Control had caused a major controversy and criticism from environmental groups that blamed the government for giving nuclear power industry control over the state EIA.

Not surprisingly the main problem concerning State EIA in Russia is the clear differentiation of jurisdiction between the two above-mentioned Federal bodies.

Sri Lanka

Environmental Impact Assessments: One popular approach to assist in smart growth in democratic countries is for law-makers to require prospective developers to prepare environmental impact assessments of their plans as a condition for state and/or local governments to go for Environmental Impact Assessments

These reports often indicate how significant impacts generated by the development will be mitigated - the cost of which is usually paid by the developer. These assessments are frequently controversial. Conservationists, neighborhood advocacy groups and NIMBYs are often skeptical about such impact reports, even when they are prepared by independent agencies and subsequently approved by the decision makers rather than the promoters. Conversely, developers will sometimes strongly resist being required to implement the mitigation measures required by the local government as they may be quite costly.

These assessments are frequently controversial. Conservationists, neighborhood advocacy groups and NIMBYs are often skeptical about such impact reports, even when they are prepared by independent agencies and subsequently approved by the decision makers rather than the promoters. Conversely, developers will sometimes strongly resist being required to implement the mitigation measures required by the local government as they may be quite costly.

The importance of the Environmental Impact Assessment as an effective tool for the purpose of integrating environmental considerations with development planning is highly recognized in Sri Lanka. The application of this technique is considered as a means of ensuring that the likely effects of new development projects on the environment are fully understood and taken into account before development is allowed to proceed. The importance of this management tool to foresee potential environmental impacts and problems caused by proposed projects and its use as a mean to make project more suitable to the environment are highly appreciated.

United States

Under United States environmental law an Environmental Assessment (EA) is compiled to determine the need for an *Environmental Impact Statement* (EIS), and originated in the National Environmental Policy Act (NEPA), enacted in 1969. Certain actions of federal agencies must be preceded by an EA or EIS. Contrary to a widespread misconception, NEPA does not prohibit the federal government or its licensees/permittees from harming the environment, nor does it specify any penalty if the EA or EIS turns out to be inaccurate, intentionally or otherwise. NEPA requires that plausible statements as to the prospective impacts be disclosed in advance. The purpose of NEPA process is to ensure that the decision maker is fully informed of the environmental aspects and consequences prior to making the final decision.

An *Environmental Assessment (EA)* is an environmental analysis prepared pursuant to the National Environmental Policy Act to determine whether a federal action would significantly affect the environment and thus require a more detailed *Environmental Impact Statement (EIS)*. The release of an Environmental Assessment results in either a *Finding of No Significant Impact (FONSI)* or an *Environmental Impact Statement (EIS)*.

The Environmental Assessment is a concise public document prepared by the federal action agency that serves to:

1. briefly provide sufficient evidence and analysis for determining whether to prepare an EIS or a Finding of No Significant Impact (FONSI)
2. Demonstrate compliance with the act when no EIS is required
3. facilitate the preparation of a EIS when a FONSI cannot be demonstrated

The Environmental Assessment includes a brief discussion of the purpose and need of the proposal and of its alternatives as required by CFR 102(2)(E), of the environmental impacts of the proposed actions and alternatives, as well as a listing of agencies and stakeholders consulted. The action agency must approve an EA before it is made available to the public. The EA is made public through notices of availability by local, state, or regional clearing houses, newspapers, etc. There is a 15-30 day review period required for an Environmental Assessment, while the document is made available for public commentary. An agency will release either a *Draft Environmental Assessment* (Draft EA) or a *Draft Environmental Impact Statement* (DEIS) for comment. Interested parties and the general public have the opportunity to comment on the draft, after which the agency will address all comments received and prepare a decision document, either a FONSI, Notice of Intent (NOI) to prepare a EIS or a Record of Decision for an EIS. The agency will then approve the " Final Environmental Assessment" (Final EA) or *Final Environmental Impact Statement* (FEIS). Commenting on the Draft EA is typically done in writing or email, submitted to the lead agency as defined in the Notice of Availability. Draft EIS's require public hearings, so comments can be made in person, as well as in writing. Occasionally, the agency will later release a "Supplemental Environmental Assessment "

(Supplemental EA) or a *Supplemental Environmental Impact Statement* (SEIS), if the project parameters or environmental conditions change substantially after the issuance of the FONSI or ROD.

The adequacy of an EIS can be challenged in federal court. Major proposed projects have been blocked because of an agency's failure to prepare an acceptable EIS. One prominent example was the Westway landfill and highway development in and along the Hudson River in New York City. Another prominent case involved the Sierra Club suing the Nevada Department of Transportation over its denial of Sierra Club's request to issue a supplemental EIS addressing air emissions of particulate matter and hazardous air pollutants in the case of widening US Highway 95 through Las Vegas. The case reached the United States Court of Appeals for the Ninth Circuit, which led to construction on the highway being halted until the court's final decision. The case was settled prior to the court's final decision.

Several state governments that have adopted "little NEPAs," state laws imposing EIS requirements for particular state actions. Some those state laws such as ()the California Environmental Quality Act refer to the required environmental impact studies as environmental impact reports.

The structure of a generic Environmental Assessment is as follows:

1. Summary
2. Introduction
    - o Structure
    - o Background
    - o Purpose and Need for Action
    - o Proposed Action
    - o Decision Framework
    - o Public Involvement
    - o Issues
3. Alternatives, including the Proposed Action
    - o Alternatives
    - o Mitigation Common to All Alternatives
    - o Comparison of Alternatives
4. Environmental Consequences
5. Consultation and Coordination

These variety of state requirements are yielding voluminous data not just upon impacts of individual projects, but also to elucidate scientific areas that had not been sufficiently researched. For example, in a seemingly routine *Environmental Impact Report* for the city of Monterey, California, information came to light that led to the official federal endangered species listing of Hickman's potentilla, a rare coastal wildflower.

Transboundary Application

Environmental threats do not respect national borders. International pollution can have detrimental effects on the atmosphere, oceans, rivers, aquifers, farmland, the weather and biodiversity. Global climate change is transnational. Specific pollution threats include acid rain, radioactive contamination, debris in outer space, stratospheric ozone depletion and toxic oil spills. The Chernobyl disaster, precipitated by a nuclear accident on April 26, 1986, is a stark reminder of the devastating effects of transboundary nuclear pollution.

Environmental protection is inherently a cross-border issue and has led to the creation of transnational regulation via multilateral and bilateral treaties. The United Nations Conference on the Human Environment (UNCHE or Stockholm Conference) held in Stockholm in 1972 and the United Nations Conference on the Environment and Development (UNCED or Rio Summit, Rio Conference, or Earth Summit) held in Rio de Janeiro in 1992 were key in the creation of about 1,000 international instruments that include at least some provisions related to the environment and its protection.

The United Nations Economic Commission for Europe's Convention on Environmental Impact Assessment in a Transboundary Context was negotiated to provide an international legal framework for transboundary EIA. However, as there is no universal legislature or administration with a comprehensive mandate, most international treaties exist parallel to one another and are further developed without the benefit of consideration being given to potential conflicts with other agreements. There is also the issue of international enforcement. This has led to duplications and failures, in part due to an inability to enforce agreements. An example is the failure of many international fisheries regimes to restrict harvesting practises.

Annexed Projects

All projects are either classified as Annex 1 or Annex 2 projects. Those lying in Annex 1 are large scale developments such as motorways, chemical works, bridges, powerstations etc. These always require an EIA under the Environmental Impact Assessment Directive (85,337,EEC). Annex 2 projects are smaller in scale such as small installments, and maintenance works. These do not require an EIA.

Environmental Management

Environmental resource management is "a purposeful activity with the goal to maintain and improve the state of an environmental resource affected by human activities". It is not, as the phrase suggests, the management of the *environment* as such, but rather the management of the interaction and impact of human societies on the environment. Environmental resources management aims to ensure that ecosystem services are protected and maintained for equitable use by future human generations, and also, maintain ecosystem integrity as an end in itself by taking into consideration ethical, economic, and scientific (ecological) variables. Environmental resource management tries to identify the factors that have a stake in the conflicts that may rise between meeting the needs and protecting the resources.

Significance

Environmental resource management is an issue of increasing concern as reflected in its prevalence in seminal texts influencing global socio-political frameworks such as the Brundtland Commission's Our Common Future which brought to the fore the integrated nature of environment and international development and the Worldwatch Institute's annual State of the World (book series) reports.

Scope

Environmental resource management can be viewed from a variety of perspectives. Environmental resource management involves the management of all components of the biophysical environment, both living (biotic) and non-living (abiotic). This is due to the interconnected and network of relationships amongst all living species and their habitats. The environment also involves the relationships of the human environment, such as the social, cultural and economic environment with the biophysical environment. The essential aspects of environmental resource management are ethical, economical, social and technological which provide for formulation of principles and help in making decisions. The scientific and technical nature makes environmental resource management profession to operate in a humanistic and rational mode in the world.

Figure: *Improved agricultural practices such as these terraces in northwest Iowa can serve to preserve soil and improve water quality*

Aspects

Ethical: Environmental resource management strategies are intrinsically driven by conceptions of human-nature relationships. Ethical aspects involve the cultural and social issues relating to the environment, and dealing with changes to it. "All human activities take place in the context of certain types of relationships between society and the bio-physical world (the rest of nature)", and so, there is a great significance in understanding the ethical values of different groups around the world. Broadly speaking, two schools of thought exist in environmental ethics: Anthropocentrism and Ecocentrism each influencing a broad spectrum of environmental resource management styles along a continuum. These styles perceive "different evidence, imperatives, and problems, and prescribe different solutions, strategies, technologies, roles for economic sectors, culture, governments, and ethics, etc".

Anthropocentrism

Anthropocentrism, "an inclination to evaluate reality exclusively in terms of human values", is an ethic reflected in the major interpretations of Western religions and the dominant economic paradigms of the industrialised world. Anthropocentrism supports an understanding of nature as existing solely for the benefit of man and as a commodity to be used for the good of humanity and improved human quality of life. Anthropocentric environmental resource management is therefore not the conservation of the environment solely for the environment's sake, but rather the conservation of the environment, and ecosystem structure, for humankind's sake.

Ecocentrism

Ecocentrists believe in the intrinsic value of nature while maintaining an understanding that "human beings must use and even exploit nature to survive and live". It is this fine ethical line that ecocentrists navigate between "fair use and downright abuse". At an extreme end of the ethical scale, ecocentrism includes philosophies such as ecofeminism and deep ecology which evolved as a reaction to the dominant anthropocentric paradigms. "In its current form, it is an attempt to synthesize many old and some new philosophical attitudes about the relationship between nature and human activity, with particular emphasis on ethical, social, and spiritual aspects that have been downplayed in the dominant economic worldview".

Economic

Figure: *A water harvesting system collects rainwater from the Rock of Gibraltar into pipes which lead to tanks excavated inside the rock.*

The economy functions within, and is dependent upon goods and services provided by natural ecosystems. The role of the environment is recognized in both classical economics and neoclassical economics theories, yet the environment held a spot on the back-burner of economic policies from 1950 – 1980 due to emphasis from policy makers on economic growth. With the prevalence of environmental problems, many economists embraced the notion that "if environmental sustainability must coexist for economic sustainability, then the overall system must be one which permits the identification of an equilibrium between the environment and the economy". As such, economic policy makers began to incorporate the functions of the natural environment – or natural capital - particularly as a sink for wastes and for the provision of raw materials and amenities.

Debate continues among economists as to how to account for natural capital, specifically whether resources can be replaced through the use of knowledge and technology, or whether the economy is a closed system that cannot be replenished and is finite. Economic models influence environmental resource management in that management policies reflect beliefs about natural capital scarcity – if natural capital is believed to be infinite and easily substituted, environmental management would be irrelevant to the economy. For example, economic paradigms based on neoclassical models of closed economic systems are primarily concerned with resource scarcity, and thus prescribe legalizing the environment as an economic externality for an environmental resource management strategy. This approach has often been termed 'Command-and-control'. Colby has identified trends in the development of economic paradigms, among them, a shift towards more ecological economics since the 1990's.

Ecological

"The pairing of significant uncertainty about the behaviour and response of ecological systems with urgent calls for near-term action constitutes a difficult reality, and a common lament" for many environmental resource managers. Scientific analysis of the environment deals with several dimensions of ecological uncertainty. These include: *structural uncertainty* resulting from the misidentification, or lack of information pertaining to the relationships between ecological variables; *parameter uncertainty* referring to "uncertainty associated with parameter values that are not known precisely but can be assessed and reported in terms of the likelihood...of experiencing a defined range of outcomes"; and *stochastic uncertainty* stemming from chance or unrelated factors. Adaptive management is considered a useful framework through which to deal with situations of high levels of uncertainty though it is not without its detractors.

A common scientific concept and impetus behind environmental resource management is carrying capacity. Simply put, carrying capacity refers to the maximum number of organisms a particular resource can sustain. The concept of carrying capacity, whilst understood by many cultures over history, has its roots in Malthusian theory. An example is visible in the EU Water Framework Directive.

However, "it is argued that Western scientific knowledge ... is often insufficient to deal with the full complexity of the interplay of variables in environmental resource management. These concerns have been recently addressed by a shift in environmental resource management approaches to incorporate different knowledge systems including traditional knowledge, reflected in approaches such as adaptive co-management community-based natural resource management and transitions management among others.

Figure:*Fencing separates big game from vehicles along the Quebec Autoroute 73 in Canada.*

Sustainability

Sustainability and environmental resource management involves managing economic, social, and ecological systems within and external to an organizational entity in order for it to sustain itself and the system it exists within. In context, sustainability implies that rather than competing for endless growth on a finite planet, development will improve quality of life without necessarily having to consume more resources. In order to sustainably manage the state of environmental resources affected by human activities organizational change is needed to instill sustainability values within an organization, in order to portray these values outwardly from all levels and to reinforce them in its surrounding stakeholder community. The end result should be a symbiotic relationship between the sustaining organization and community, along with the environment.

There are many drivers that compel environmental resource management to take sustainability issues into account. Today's economic paradigms do not protect the natural environment, yet they deepen human dependency on biodiversity and ecosystem services. Ecologically, massive environmental degradation and climate change threaten the stability of ecological systems that humanity depends on. Socially, an increasing gap between rich and poor and the global North-South divide denies many access to basic human needs, rights, and education, leading to further environmental destruction. The planet's unstable condition is caused by many anthropogenic sources. As an exceptionally powerful contributing factor to social and environmental change, the modern organisation has the potential to apply environmental resource management with sustainability principals to achieve highly affective outcomes. To achieve sustainable development with environmental resource management an organisation should coincide with sustainability principals, such as: social and environmental accountability, long-term planning; a strong, shared vision; a holistic focus; devolved and consensus decision making; broad stakeholder engagement and justice; transparency measures; trust; and flexibility, to name a few.

Current Paradigm Shifts

In order to adjust to today's environment of quick social and ecological changes some organizations have begun to experiment with various new tools and concepts. Those which are more traditional and stick to hierarchal decision making are having difficulty dealing with the demand for lateral decision making that supports affective participation. Whether it be a matter of ethics or just strategic advantage organizations are internalizing sustainability principles. Examples of some of the world's largest and most profitable corporations who are shifting to sustainable environmental resource management are: Ford, Toyota, BMW, Honda, Shell, Du Pont, Swiss Re, Hewlett-Packard, and Unilever. An extensive study by the Boston Consulting Group reaching 1,560 business leaders from diverse regions, job positions, expertise in sustainability, industries, and sizes of organizations, revealed the many benefits of sustainable practice as well as its viability.

It is important to note that though sustainability of environmental resource management has improved, corporate sustainability, for one, has yet to reach the majority of global companies operating in the markets. The three major barriers to preventing organizations to shift towards sustainable practice with environmental resource management are: not understanding what sustainability is; having difficulty modeling an economically viable case for the switch; and having a flawed execution plan, or a lack there of. Therefore the most important part of shifting an organization to adopt sustainability in environmental resource management would be to create a shared vision and understanding of what sustainability is for that particular organization, and to clarify the business case.

Stakeholders

Public Sector: The public sector comprises the general government sector plus all public corporations including the central bank. In environmental resource management the public sector is responsible for administering natural resource management and implementing environmental protection legislation. The traditional role of the public sector in environmental resource management is to provide professional judgement through skilled technicians on behalf of the public. With the increase of intractable environmental problems, the public sector has been led to examine alternative paradigms for managing environmental resources. This has resulted in the public sector working collaboratively with other sectors (including other governments, private and civil) to encourage sustainable natural resource management behaviours.

Figure: *A conservation project in North Carolina involving the search for bog turtles was conducted by United States Fish and Wildlife Service and the North Carolina Wildlife Resources Commission and its volunteers*

Private Sector

The private sector comprises private corporations and non-profit institutions serving households. The private sector's traditional role in environmental resource management is that of the recovers of natural resources. Such private sector recovery groups include mining (minerals and petroleum), forestry and fishery organisations. Environmental resource management undertaken by the private sectors varies dependent upon the resource type, that being renewable or non-renewable and private and common resources. Environmental managers from the private sector also need skills to manage collaboration within a dynamic social and political environment.

Civil Society

Civil society comprises associations in which societies voluntarily organise themselves into and which represent a wide range of interests and ties. These can include community-based organisations, indigenous peoples' organisations and non-government organisations (NGO). Functioning through strong public pressure, civil society can exercise their legal rights against the implementation of resource management plans, particularly land management plans. The aim of civil society in environmental resource management is to be included in the decision-making process by means of public participation. Public participation can be an effective strategy to invoke a sense of social responsibility of natural resources.

Tools

As with all management functions, effective management tools, standards and systems are required. An environmental management standard or system or protocol attempts to reduce environmental impact as measured by some objective criteria. The ISO 14001 standard is the most widely used standard for environmental risk management and is closely aligned to the European Eco-Management and Audit Scheme (EMAS). As a common auditing standard, the ISO 19011 standard explains how to combine this with quality management.

Other environmental management systems (EMS) tend to be based on the ISO 14001 standard and many extend it in various ways:

- The Green Dragon Environmental Management Standard is a five level EMS designed for smaller organisations for whom ISO 14001 may be too onerous and for larger organisations who wish to implement ISO 14001 in a more manageable step-by-step approach,
- BS 8555 is a phased standard that can help smaller companies move to ISO 14001 in six manageable steps,
- The Natural Step focuses on basic sustainability criteria and helps focus engineering on reducing use of materials or energy use that is unsustainable in the long term,

- Natural Capitalism advises using accounting reform and a general biomimicry and industrial ecology approach to do the same thing,
- US Environmental Protection Agency has many further terms and standards that it defines as appropriate to large-scale EMS,
- The UN and World Bank has encouraged adopting a "natural capital" measurement and management framework,
- The European Union Eco-Management and Audit Scheme (EMAS).

Other strategies exist that rely on making simple distinctions rather than building top-down management "systems" using performance audits and full cost accounting. For instance, Ecological Intelligent Design divides products into consumables, service products or durables and unsaleables - toxic products that no one should buy, or in many cases, do not realize they are buying. By eliminating the unsaleables from the comprehensive outcome of any purchase, better environmental resource management is achieved without "systems".

Recent successful cases have put forward the notion of "Integrated Management". It shares a wider approach and stresses out the importance of interdisciplinary assessment. It is an interesting notion that might not be adaptable to all cases.

Environmental Restoration

Environmental restoration is a term common in the citizens' environmental movement. Environmental restoration is closely allied with (or perhaps sometimes used interchangeably with) *ecological restoration* or *environmental remediation.* In the U.S., remediation is the term used more in the realms of industry, public policy, and the civil services.

In the 1987 edition of his book *Restoring the Earth: How Americans are Working to Renew our Damaged Environment*, scientific editor and writer John J. Berger defined environmental restoration (or "natural resource restoration") as follows: *"... A process in which a damaged resource is renewed. Biologically. Structurally. Functionally."*

11

Design of Grit Chambers

Primary Sedimentation

Primary sedimentation in a municipal wastewater treatment plant is generally plain sedimentation without the use of chemicals. In treating certain industrial wastes chemically aided sedimentation may be involved. In either case, it constitutes *flocculent settling,* and the particles do not remain discrete as in the case of grit, but tend to agglomerate or coagulate during settling. Thus, their diameter keeps increasing and settlement proceeds at an over increasing velocity. Consequently, they trace a curved profile. The settling tank design in such cases depends on both *surface loading* and *detention time.* Long tube settling tests can be performed in order to estimate specific value of surface loading and detention time for desired efficiency of clarification for a given industrial wastewater using recommended methods of testing. Scale-up factors used in this case range from 1.25 to 1.75 for the overflow rate, and from 1.5 to 2.0 for detention time when converting laboratory results to the prototype design.

For primary settling tanks treating municipal or domestic sewage, laboratory tests are generally not necessary, and recommended design values given in table may be used. Using an appropriate value of surface loading from table, the required tank area is computed. Knowing the average depth, the detention time is then computed. Excessively high detention time (longer than 2.5 h) must be avoided especially in warm climates where anaerobicity can be quickly induced.

Design Parameters for Settling Tank

1. ***Nutritional Requirements*:** On the basis of chemical form of carbon required, microorganisms are classified as

a. Autotrophic: organisms that use CO_2 or HCO_3^- as their sole source of carbon.

***Table** : Classification of Micro organisms*

Types of settling	*Overflow rate m^3m^2/day*		*Solids loading $kg/m^2/day$*		*Depth*	*Detention time*
	Average	Peak	Average	Peak		
Primary settling only	25-30	50-60	-	-	2.5-3.5	2.0-2.5
Primary settling followed by secondary treatment	35-50	60-120	-	-	2.5-3.5	
Primary settling with activated sludge return	25-35	50-60	-	-	3.5-4.5	-
Secondary settling for trickling filters	15-25	40-50	70-120	190	2.5-3.5	1.5-2.0
Secondary settling for activated sludge (excluding extended aeration)	15-35	40-50	70-140	210	3.5-4.5	-
Secondary settling for extended aeration	8-15	25-35	25-120	170	3.5-4.5	-

b. Heterotrophic: organisms that use carbon from organic compounds.

2. *Energy Requirements*: On the basis of energy source required, microorganisms are classified as

   Phototrophs: organisms that use light as their energy source.

   a. Chemotrophs: organisms that employ oxidation-reduction reactions to provide energy. They are further classified on the basis of chemical compounds oxidized (i.e., electron donor)

      i. Chemoorganotrophs: Organisms that use complex organic molecules as their electron donor.

      ii. Chemoautotrophs: Organisms that use simple inorganic molecules such as hydrogen sulfide or ammonia as their electron donor.

3. *Temperature Range*: On the basis of temperature range within which they can proliferate, microorganisms are classified as

   Psychrophilic: organisms whose growth is optimum within 15 to 30°C.

   a. *Mesophilic:* organisms whose growth is optimum within 30 to 45°C.

   b. *Thermophilic:* organisms whose growth is optimum within 45 to 70°C.

4. *Oxygen Requirements*: On the basis of oxygen requirement microorganisms are classified as

Aerobes: organisms that use molecular oxygen as electron acceptor.

a. Anaerobes: organisms that use some molecule other than molecular oxygen as electron acceptor.

b. Facultative organisms : organisms that can use either molecular oxygen or some other chemical compound as electron acceptor.

Growth Pattern of Micro Organisms

When a small number of viable bacterial cells are placed in a close vessel containing excessive food supply in a suitable environment, conditions are established in which unrestricted growth takes place. However, growth of an organism do not go on indefinitely, and after a characteristic size is reached, the cell divides due to hereditary and internal limitations.

The curve shown may be divided into six well defined phases:

1. *Lag Phase*:adaptation to new environment, long generation time and null growth rate.
2. *Accelaration Phase*: decreasing generation time and increasing growth rate.
3. *Exponential Phase*: minimal and constant generation time, maximal and constant specific growth rate and maximum rate of substrate conversion.
4. *Declining Growth Phase*: increasing generation time and decreasing specific growth rate due to gradual decrease in substrate concentration and increased accumulation of toxic metabolites.
5. *Stationary Phase*: exaustion of nutrients, high concentration of toxic metabolites, and cells in a state of suspended animation.
6. *Endogenous Phase*: endogenous metabolism, high death rate and cell lysis.

Biomass Growth Rate

The most widely used expression for the growth rate of micro organisms is given by Monod:

Total rate of microbial growth, $\frac{dx}{dt} = \frac{m_m XS}{K_s + S}$

where,

m_m = maximum specific growth rate

X = micro organism concentration

S = substrate concentration

K_s = substrate concentration at one half the maximum growth rate

Similarly, rate of substrate utilization,

$$\frac{dS}{dt} = \frac{k X S}{K_s + S}$$

where,

k = maximum specific substrate utilization rate

Maintenance as Endogenous Respiration

Net growth rate of micro organisms is computed by subtracting from the total growth rate, the rate of micro organisms endogenously decayed to satisfy maintenance energy requirement. Therefore,

$$\text{Net rate of microbial growth} = \frac{m_m X S}{K_s + S} - k_d X$$

where,

k_d = endogenous decay coefficient

Growth Yield

Growth yield is defined as the incremental increase in biomass which results from the utilization of the incremental amount of substrate. The maximum specific growth rate is given by: m_m =Y.k

where, Y is the maximum yield coefficient and is defined as the ratio of maximum mass of cells formed to the mass of substrate utilized. The coefficients Y, k_d, k and K_s are designated as kinetic coefficients. The values of kinetic coefficients depend upon the nature of wastewater and operational and environmental conditions in biological reactor. The biological reactors can be completely mixed flow or plug flow reactor with or without recycle.

Activated Sludge Process

In activated sludge process wastewater containing organic matter is aerated in an aeration basin in which micro-organisms metabolize the suspended and soluble organic matter. Part of organic matter is synthesized into new cells and part is oxidized to CO_2 and water to derive energy. In activated sludge systems the new cells formed in the reaction are removed from the liquid stream in the form of a flocculent sludge in settling tanks. A part of this settled biomass, described as activated sludge is returned to the aeration tank and the remaining forms waste or excess sludge.

Activated Sludge Process Variables

The main variables of activated sludge process are the mixing regime, loading rate, and the flow scheme.

Mixing Regime

Generally two types of mixing regimes are of major interest in activated sludge process: *plug flow* and *complete mixing*. In the first one, the regime is characterized by orderly flow of mixed liquor through the aeration tank with no element of mixed liquor overtaking or mixing with any other element. There may be lateral mixing of mixed liquor but there must be no mixing along the path of flow.In complete mixing, the contents of aeration tank are well stirred and uniform throughout. Thus, at steady state, the effluent from the aeration tank has the same composition as the aeration tank contents.

The type of mixing regime is very important as it affects (1) oxygen transfer requirements in the aeration tank, (2) susceptibility of biomass to shock loads, (3) local environmental conditions in the aeration tank, and (4) the kinetics governing the treatment process.

Design Consideration

The items for consideration in the design of activated sludge plant are aeration tank capacity and dimensions, aeration facilities, secondary sludge settling and recycle and excess sludge wasting.

Oxygen Requirements

Oxygen is reqiured in the activated sludge process for the oxidation of a part of the influent organic matter and also for the endogenous respiration of the micro-organisms in the system. The formula does not allow for nitrification but allows only for carbonaceous BOD removal.

Aeration Facilities

The aeration facilities of the activated sludge plant are designed to provide the calculated oxygen demand of the wastewater against a specific level of dissolved oxygen in the wastewater.

Secondary Settling

Secondary settling tanks, which receive the biologically treated flow undergo zone or compression settling. *Zone settling* occurs beyond a certain concentration when the particles are close enough together that interparticulate forces may hold the particles fixed relative to one another so that the whole mass tends to settle as a single layer or "blanket" of sludge. The rate at which a sludge blanket settles can be determined by timing its position in a settling column test whose results can be plotted as shown in figure.

Compression settling may occur at the bottom of a tank if particles are in such a concentration as to be in physical contact with one another. The weight of particles is partly supported by the lower layers of particles, leading to progressively greater compression with depth and thickening of sludge. From the settling column test, the limiting solids flux required to reach any desired underflow concentration can be estimated, from which the rquired tank area can be computed.

The solids load on the clarifier is estimated in terms of (Q+R)X, while the overflow rate or surface loading is estimated in terms of flow Q only (not Q+R) since the quantity R is withdrawn from the bottom and does not contribute to the overflow from the tank. The secondary settling tank is particularly sensitive to fluctuations in flow rate and on this account it is recommended that the units be designed not only for average overflow rate but also for peak overflow rates. Beyond an MLSS concentration of 2000 mg/l the clarifier design is often controlled by the solids loading rate rather than the overflow rate. Recommended design values for treating domestic sewage in final clarifiers and mechanical thickeners (which also fall in this category of compression settling) are given in lecture 22.

Sludge Recycle

The sludge settleability is determined by sludge volume index (SVI) defined as volume occupied in mL by one gram of solids in the mixed liquor after settling for 30 min. If it is assumed that sedimentation of suspended solids in the laboratory is similar to that in sedimentation tank, then $X_r = 10^6/SVI$. Values of SVI between 100 and 150 ml/g indicate good settling of suspended solids. The X_r value may not be taken more than 10,000 g/m^3 unless separate thickeners are provided to concentrate the settled solids or secondary sedimentation tank is designed to yield a higher value.

Excess Sludge Wasting

The sludge in the aeration tank has to be wasted to maintain a steady level of MLSS in the system. The excess sludge quantity will increase with increasing F/M and decrease with increasing temperature. Excess sludge may be wasted either from the sludge return line or directly from the aeration tank as mixed liquor.

Pumps

A pump is a device used to move fluids, such as liquids, gases or slurries. A pump displaces a volume by physical or mechanical action. Pumps fall into three major groups: direct lift, displacement, and gravity pumps. Their names describe the method for moving a fluid.

Types

Positive Displacement Pump: A positive displacement pump causes a fluid to move trapping a fixed amount of it then forcing (displacing) that trapped volume into the discharge pipe. Some positive displacement pumps work using an expanding cavity on the suction side and a decreasing cavity on the discharge side. Liquid flows into the pump as the cavity on the suction side expands and the liquid flows out of the discharge as the cavity collapses. The volume is constant given each cycle of operation.

Positive Displacement Pump Behavior and Safety

Positive displacement pumps, unlike centrifugal or roto-dynamic pumps, will produce the same flow at a given speed (RPM) no matter what the discharge pressure. Thus, positive displacement pumps are "constant flow machines". A positive displacement pump must not be operated against a closed valve on the discharge side of the pump, because it has no shut-off head like centrifugal pumps. A positive displacement pump operating against a closed discharge valve will continue to produce flow and the pressure in the discharge line will increase, until the line bursts or the pump is severely damaged, or both. A relief or safety valve on the discharge side of the positive displacement pump is therefore necessary. The relief valve can be internal or external. The pump manufacturer normally has the option to supply internal relief or safety valves. The internal valve should in general only be used as a safety precaution, an external relief valve installed in the discharge line with a return line back to the suction line or supply tank is recommended.

Positive Displacement Types

A positive displacement pump can be further classified according to the mechanism used to move the fluid:

- Rotary-type positive displacement: internal gear, screw, shuttle block, flexible vane or sliding vane, circumferential piston, helical twisted roots (e.g. the Wendelkolben pump) or liquid ring vacuum pumps.
- Reciprocating-type positive displacement: piston or diaphragm pumps.
- Linear-type positive displacement: Rope pumps and chain pumps.

Rotary Positive Displacement Pumps

Positive displacement rotary pumps are pumps that move fluid using the principles of rotation. The vacuum created by the rotation of the pump captures and draws in the liquid. Advantages: Rotary pumps are very efficient because they naturally remove air from the lines, eliminating the need to bleed the air from the lines manually.

Drawbacks: Positive displacement rotary pumps also have their weaknesses. Because of the nature of the pump, the clearance between the rotating pump and the outer edge must be very close, requiring that the pumps rotate at a slow, steady speed. If rotary pumps are operated at high speeds, the fluids will cause erosion. Rotary pumps that experience such erosion eventually show signs of enlarged clearances, which allow liquid to slip through and reduce the efficiency of the pump. Rotary positive displacement pumps can be grouped into three main types:

- Gear pumps - a simple type of rotary pump where the liquid is pushed between two gears.
- Screw pumps - the shape of the internals of this pump usually two screws turning against each other pump the liquid.
- Rotary vane pumps - similar to scroll compressors, consisting of a cylindrical rotor encased in a similarly shaped housing. As the rotor turns, the vanes trap fluid between the rotor and the casing, drawing the fluid through the pump.

Reciprocating Positive Displacement Pumps

Reciprocating pumps are those which cause the fluid to move using one or more oscillating pistons, plungers or membranes (diaphragms).

Reciprocating-type pumps require a system of suction and discharge valves to ensure that the fluid moves in a positive direction. Pumps in this category range from having "simplex" one cylinder, to in some cases "quad" four cylinders or more. Most reciprocating-type pumps are "duplex" (two) or "triplex" (three) cylinder. Furthermore, they can be either "single acting" independent suction and discharge strokes or "double acting" suction and discharge in both directions. The pumps can be powered by air, steam or through a belt drive from an engine or motor. This type of pump was used extensively in the early days of steam propulsion (19th century) as boiler feed water pumps. Reciprocating pumps are now typically used for pumping highly viscous fluids including concrete and heavy oils, and special applications demanding low flow rates against high resistance.

Figure: *Hand-operated, reciprocating, positive displacement, water pump in Košice-ahanovce, Slovakia (walking beam pump).*

These positive displacement pumps have an expanding cavity on the suction side and a decreasing cavity on the discharge side. Liquid flows into the pumps as the cavity on the suction side expands and the liquid flows out of the discharge as the cavity collapses. The volume is constant given each cycle of operation.

Typical reciprocating pumps are:

- plunger pumps - a reciprocating plunger pushes the fluid through one or two open valves, closed by suction on the way back.
- diaphragm pumps - similar to plunger pumps, where the plunger pressurizes hydraulic oil which is used to flex a diaphragm in the pumping cylinder. Diaphragm valves are used to pump hazardous and toxic fluids.
- piston displacement pumps - usually simple devices for pumping small amounts of liquid or gel manually. An example is the common hand soap pump.

Various Positive Displacement Pumps

The positive displacement principle applies in the following types of pumps:

- rotary lobe pump
- Progressive cavity pump
- rotary gear pump
- piston pump
- diaphragm pump
- screw pump
- gear pump
- Hydraulic pump
- vane pump
- regenerative (peripheral) pump
- peristaltic pump
- rope pump

Gear Pump

This is the simplest of rotary positive displacement pumps. It consists of two meshed gears rotating in a closely fitted casing. Fluid is pumped around the outer periphery by being trapped in the tooth spaces. It does not travel back on the meshed part, since the teeth mesh closely in the centre. It is widely used on car engine oil pumps, and also in various hydraulic power packs.

Screw Pump

Screw pumps are a more complicated type of rotary pumps, featuring two or three screws with opposing thread —- that is, one screw turns clockwise, and the other counterclockwise. The screws are each mounted on shafts that run parallel to each other; the shafts also have gears on them that mesh with each other in order to turn the shafts together and keep everything in place. The turning of the screws, and consequently the shafts to which they are mounted, draws the fluid through the pump. As with other forms of rotary pumps, the clearance between moving parts and the pump's casing is minimal.

Progressing cavity pump

Widely used for pumping difficult materials such as sewage sludge contaminated with large particles, this pump consists of a helical shaped rotor, about ten times as long as its width. This can be visualized as a central core of diameter x, with typically a curved spiral wound around of thickness half x, although of course in reality it is made from one casting. This shaft fits inside a heavy duty rubber sleeve, of wall thickness typically x also. As the shaft rotates, fluid is gradually forced up the rubber sleeve. Such pumps can develop very high pressure at quite low volumes.

Roots-type pumps

Named after the Roots brothers who designed and invented it, this lobe pump works by displacing the liquid trapped between two long helical twisted rotors, each fitting into the other when perpendicular at 90°, rotating inside a triangular shaped sealing line configuration, both at the point of suction and at the point of discharge.

This design produces a continuous flow with equal volume and no vortex. It can work at low pulsation rates and results with gentle performance, more fit for some applications.

Some applications are:

- High capacity industrial air compressors
- Roots Type Superchargers on internal combustion engines.
- A brand of civil defense siren, the Federal Signal Corporation's Thunderbolt.

Peristaltic Pump

A peristaltic pump is a type of positive displacement pump used for pumping a variety of fluids. The fluid is contained within a flexible tube fitted inside a circular pump casing (though linear peristaltic pumps have been made). A rotor with a number of "rollers", "shoes" or "wipers" attached to the external circumference compresses the flexible tube. As the rotor turns, the part of the tube under compression closes (or "occludes") thus forcing the fluid to be pumped to move through the tube. Additionally, as the tube opens to its natural state after the passing of the cam ("restitution") fluid flow is induced to the pump. This process is called peristalsis and is used in many biological systems such as the gastrointestinal tract.

Plunger Pumps

Plunger pumps are reciprocating positive displacement pumps. They consist of a cylinder with a reciprocating plunger in them. The suction and discharge valves are mounted in the head of the cylinder. In the suction stroke the plunger retracts and the suction valves open causing suction of fluid into the cylinder. In the forward stroke the plunger pushes the liquid out of the discharge valve.

Efficiency and common problems: With only one cylinder in plunger pumps, the fluid flow varies between maximum flow when the plunger moves through the middle positions, and zero flow when the plunger is at the end positions. A lot of energy is wasted when the fluid is accelerated in the piping system. Vibration and "water hammer" may be a serious problem. In general the problems are compensated for by using two or more cylinders not working in phase with each other.

Compressed-air-powered Double-diaphragm Pumps

One modern application of positive displacement diaphragm pumps is compressed-air-powered double-diaphragm pumps. Run on compressed air these pumps are intrinsically safe by design, although all manufacturers offer ATEX certified models to comply with industry regulation. Commonly seen in all areas of industry from shipping to processing, Wilden Pumps, Graco, SandPiper or ARO are generally the larger of the brands. They are relatively inexpensive and can be used for almost any duty from pumping water out of bunds, to pumping hydrochloric acid from secure storage (dependent on how the pump is manufactured – elastomers / body construction). Lift is normally limited to roughly 6m although heads can reach almost 200 Psi.

Rope Pumps

Devised in China as chain pumps over 1000 years ago, these pumps can be made from very simple materials: A rope, a wheel and a PVC pipe are sufficient to make a simple rope pump. For this reason they have become extremely popular around the world since the 1980s. Rope pump efficiency has been studied by grass roots organizations and the techniques for making and running them have been continuously improved.

Impulse Pumps

Impulse pumps use pressure created by gas (usually air). In some impulse pumps the gas trapped in the liquid (usually water), is released and accumulated somewhere in the pump, creating a pressure which can push part of the liquid upwards.

Impulse pumps include:

- Hydraulic ram pumps - uses pressure built up internally from released gas in liquid flow.

- Pulser pumps - run with natural resources, by kinetic energy only.
- Airlift pumps - run on air inserted into pipe, pushing up the water, when bubbles move upward, or on pressure inside pipe pushing water up.

Hydraulic Ram Pumps

A hydraulic ram is a water pump powered by hydropower. It functions as a hydraulic transformer that takes in water at one "hydraulic head" (pressure) and flow-rate, and outputs water at a higher hydraulic-head and lower flow-rate. The device uses the water hammer effect to develop pressure that allows a portion of the input water that powers the pump to be lifted to a point higher than where the water originally started.

The hydraulic ram is sometimes used in remote areas, where there is both a source of low-head hydropower, and a need for pumping water to a destination higher in elevation than the source. In this situation, the ram is often useful, since it requires no outside source of power other than the kinetic energy of flowing water.

Velocity Pumps

A centrifugal pump uses a spinning "impeller" which has backward-swept arms. Rotodynamic pumps (or dynamic pumps) are a type of velocity pump in which kinetic energy is added to the fluid by increasing the flow velocity. This increase in energy is converted to a gain in potential energy (pressure) when the velocity is reduced prior to or as the flow exits the pump into the discharge pipe. This conversion of kinetic energy to pressure can be explained by the First law of thermodynamics or more specifically by Bernoulli's principle. Dynamic pumps can be further subdivided according to the means in which the velocity gain is achieved.

These types of pumps have a number of characteristics:

1. Continuous energy
2. Conversion of added energy to increase in kinetic energy (increase in velocity)
3. Conversion of increased velocity (kinetic energy) to an increase in pressure head

One practical difference between dynamic and positive displacement pumps is their ability to operate under closed valve conditions. Positive displacement pumps physically displace the fluid; hence closing a valve downstream of a positive displacement pump will result in a continual build up in pressure resulting in mechanical failure of either pipeline or pump. Dynamic pumps differ in that they can be safely operated under closed valve conditions (for short periods of time).

Centrifugal Pump

A centrifugal pump is a rotodynamic pump that uses a rotating impeller to increase the pressure and flow rate of a fluid. Centrifugal pumps are the most common type of pump used to move liquids through a piping system. The fluid enters the pump impeller along or near to the rotating axis and is accelerated by the impeller, flowing radially outward or axially into a diffuser or volute chamber, from where it exits into the downstream piping system. Centrifugal pumps are typically used for large discharge through smaller heads.

Centrifugal pumps are most often associated with the radial flow type. However, the term "centrifugal pump" can be used to describe all impeller type rotodynamic pumps including the radial, axial and mixed flow variations.

Radial Flow Pumps

Often simply referred to as centrifugal pumps. The fluid enters along the axial plane, is accelerated by the impeller and exits at right angles to the shaft (radially). Radial flow pumps operate at higher pressures and lower flow rates than axial and mixed flow pumps.

Axial Flow Pumps

Axial flow pumps differ from radial flow in that the fluid enters and exits along the same direction parallel to the rotating shaft. The fluid is not accelerated but instead "lifted" by the action of the impeller. They may be likened to a propeller spinning in a length of tube. Axial flow pumps operate at much lower pressures and higher flow rates than radial flow pumps.

Mixed Flow Pumps

Mixed flow pumps, as the name suggests, function as a compromise between radial and axial flow pumps, the fluid experiences both radial acceleration and lift and exits the impeller somewhere between 0–90 degrees from the axial direction. As a consequence mixed flow pumps operate at higher pressures than axial flow pumps while delivering higher discharges than radial flow pumps. The exit angle of the flow dictates the pressure head-discharge characteristic in relation to radial and mixed flow.

Eductor-jet Pump

This uses a jet, often of steam, to create a low pressure. This low pressure sucks in fluid and propels it into a higher pressure region.

Gravity Pumps

Gravity pumps include the syphon and Heron's fountain – and there also important qanat or foggara systems which simply use downhill flow to take water from far-underground aquifers in high areas to consumers at lower elevations. The hydraulic ram is also sometimes referred to as a gravity pump.

Steam Pumps

Steam pumps have been for a long time mainly of historical interest. They include any type of pump powered by a steam engine and also pistonless pumps such as Thomas Savery's, the Pulsometer steam pump or the Steam injection pump. Recently there has been a resurgence of interest in low power solar steam pumps for use in smallholder irrigation in developing countries. Previously small steam engines have not been viable because of escalating inefficiencies as vapour engines decrease in size. However the use of modern engineering materials coupled with alternative engine configurations has meant that these types of system are now a cost effective opportunity.

Valveless Pumps

Valveless pumping assists in fluid transport in various biomedical and engineering systems. In a valveless pumping system, no valves are present to regulate the flow direction. The fluid pumping efficiency of a valveless system, however, is not necessarily lower than that having valves. In fact, many fluid-dynamical systems in nature and engineering more or less rely upon valveless pumping to transport the working fluids therein. For instance, blood circulation in the cardiovascular system is maintained to some extent even when the heart's valves fail. Meanwhile, the embryonic vertebrate heart begins pumping blood long before the development of discernable chambers and valves. In microfluidics, valveless impedance pump have been fabricated, and are expected to be particularly suitable for handling sensitive biofluids.

Pump Repairs

Examining pump repair records and MTBF (mean time between failures) is of great importance to responsible and conscientious pump users. In view of that fact, the preface to the 2006 Pump User's

Handbook alludes to "pump failure" statistics. For the sake of convenience, these failure statistics often are translated into MTBF (in this case, installed life before failure).

In early 2005, Gordon Buck, John Crane Inc.'s chief engineer for Field Operations in Baton Rouge, LA, examined the repair records for a number of refinery and chemical plants to obtain meaningful reliability data for centrifugal pumps. A total of 15 operating plants having nearly 15,000 pumps were included in the survey. The smallest of these plants had about 100 pumps; several plants had over 2000. All facilities were located in the United States. In addition, considered as "new," others as "renewed" and still others as "established." Many of these plants—but not all—had an alliance arrangement with John Crane. In some cases, the alliance contract included having a John Crane Inc. technician or engineer on-site to coordinate various aspects of the program.

Not all plants are refineries, however, and different results can be expected elsewhere. In chemical plants, pumps have traditionally been "throw-away" items as chemical attack can result in limited life. Things have improved in recent years, but the somewhat restricted space available in "old" DIN and ASME-standardized stuffing boxes places limits on the type of seal that can be fitted. Unless the pump user upgrades the seal chamber, only the more compact and simple versions can be accommodated. Without this upgrading, lifetimes in chemical installations are generally believed to be around 50 to 60 percent of the refinery values.

It goes without saying that unscheduled maintenance often is one of the most significant costs of ownership, and failures of mechanical seals and bearings are among the major causes. Keep in mind the potential value of selecting pumps that cost more initially, but last much longer between repairs. The MTBF of a better pump may be one to four years longer than that of its non-upgraded counterpart. Consider that published average values of avoided pump failures range from $2600 to $12,000. This does not include lost opportunity costs. One pump fire occurs per 1000 failures. Having fewer pump failures means having fewer destructive pump fires.

As has been noted, a typical pump failure based on actual year 2002 reports, costs $5,000 on average. This includes costs for material, parts, labor and overhead. Let us now assume that the MTBF for a particular pump is 12 months and that it could be extended to 18 months. This would result in a cost avoidance of $2,500/yr—which is greater than the premium one would pay for the reliability-upgraded centrifugal pump.

Applications

Pumps are used throughout society for a variety of purposes. Early applications includes the use of the windmill or watermill to pump water. Today, the pump is used for irrigation, water supply, gasoline supply, air conditioning systems, refrigeration (usually called a compressor), chemical movement, sewage movement, flood control, marine services, etc.

Because of the wide variety of applications, pumps have a plethora of shapes and sizes: from very large to very small, from handling gas to handling liquid, from high pressure to low pressure, and from high volume to low volume.

Priming a Pump

Liquid and slurry pumps can lose prime and this will require the pump to be primed by adding liquid to the pump and inlet pipes to get the pump started. Loss of "prime" is usually due to ingestion of air into the pump. The clearances and displacement ratios in pumps used for liquids and other more viscous fluids cannot displace the air due to its lower density.

Pumps as Public Water Supplies

One sort of pump once common worldwide was a hand-powered water pump, or 'pitcher pump'. It would be installed over a community water well that was used by people in the days before piped water supplies. In parts of the British Isles, it was often called "the parish pump". Although such community pumps are no longer common, the expression "parish pump" is still used. It derives from the kind of the chatter and conversation that might be heard as people congregated to draw water from the community water pump, and is now used to describe a place or forum where matter of purely local interest is discussed. Because water from pitcher pumps is drawn directly from the soil, it is more prone to contamination. If such water is not filtered and purified, consumption of it might lead to gastrointestinal or other water-borne diseases.

Modern hand operated community pumps are considered the most sustainable low cost option for safe water supply in resource poor settings, often in rural areas in developing countries. A hand pump opens access to deeper groundwater that is often not polluted and also improves the safety of a well by protecting the water source from contaminated buckets. Pumps like the Afridev pump are designed to be cheap to build and install, and easy to maintain with simple parts. However, scarcity of spare parts for these type of pumps in some regions of Africa has diminished their utility for these areas.

Sealing Multiphase Pumping Applications

Multiphase pumping applications, also referred to as tri-phase, have grown due to increased oil drilling activity. In addition, the economics of multiphase production is attractive to upstream operations as it leads to simpler, smaller in-field installations, reduced equipment costs and improved production rates. In essence, the multiphase pump can accommodate all fluid stream properties with one piece of equipment, which has a smaller footprint. Often, two smaller multiphase pumps are installed in series rather than having just one massive pump.

For midstream and upstream operations, multiphase pumps can be located onshore or offshore and can be connected to single or multiple wellheads. Basically, multiphase pumps are used to transport the untreated flow stream produced from oil wells to downstream processes or gathering facilities. This means that the pump may handle a flow stream (well stream) from 100 percent gas to 100 percent liquid and every imaginable combination in between. The flow stream can also contain abrasives such as sand and dirt. Multiphase pumps are designed to operate under changing/fluctuating process conditions. Multiphase pumping also helps eliminate emissions of greenhouse gases as operators strive to minimize the flaring of gas and the venting of tanks where possible.

Types and Features of Multiphase Pumps

Helico-Axial Pumps (Centrifugal) A rotodynamic pump with one single shaft requiring two mechanical seals. This pump utilizes an open-type axial impeller. This pump type is often referred to as a "Poseidon Pump" and can be described as a cross between an axial compressor and a centrifugal pump.

Twin Screw (Positive Displacement) The twin screw pump is constructed of two intermeshing screws that force the movement of the pumped fluid. Twin screw pumps are often used when pumping conditions contain high gas volume fractions and fluctuating inlet conditions. Four mechanical seals are required to seal the two shafts.

Progressive Cavity Pumps (Positive Displacement) Progressive cavity pumps are single-screw types typically used in shallow wells or at the surface. This pump is mainly used on surface applications where the pumped fluid may contain a considerable amount of solids such as sand and dirt.

Electric Submersible Pumps (Centrifugal) These pumps are basically multistage centrifugal pumps and are widely used in oil well applications as a method for artificial lift. These pumps are usually specified when the pumped fluid is mainly liquid.

Buffer Tank A buffer tank is often installed upstream of the pump suction nozzle in case of a slug flow. The buffer tank breaks the energy of the liquid slug, smoothes any fluctuations in the incoming flow and acts as a sand trap. As the name indicates, multiphase pumps and their mechanical seals can encounter a large variation in service conditions such as changing process fluid composition, temperature variations, high and low operating pressures and exposure to abrasive/ erosive media. The challenge is selecting the appropriate mechanical seal arrangement and support system to ensure maximized seal life and its overall effectiveness.

Specifications

Pumps are commonly rated by horsepower, flow rate, outlet pressure in feet (or metres) of head, inlet suction in suction feet (or metres) of head. The head can be simplified as the number of feet or metres the pump can raise or lower a column of water at atmospheric pressure. From an initial design point of view, engineers often use a quantity termed the specific speed to identify the most suitable pump type for a particular combination of flow rate and head.

Pump Material

The pump material can be Stainless steel (SS 316 or SS 304), cast iron etc. It depends on the application of the pump. In the water industry and for pharma applications SS 316 is normally used, as stainless steel gives better results at high temperatures.

Pumping Power

The power imparted into a fluid will increase the energy of the fluid per unit volume. Thus the power relationship is between the conversion of the mechanical energy of the pump mechanism and the fluid elements within the pump. In general, this is governed by a series of simultaneous differential equations, known as the Navier-Stokes equations. However a more simple equation relating only the different energies in the fluid, known as Bernoulli's equation can be used. Hence the power, P, required by the pump:

$$P = \frac{\Delta PQ}{\eta}$$

where ΔP is the change in total pressure between the inlet and outlet (in Pa), and Q, the fluid flowrate is given in m^3/s. The total pressure may have gravitational, static pressure and kinetic energy components; i.e. energy is distributed between change in the fluid's gravitational potential energy (going up or down hill), change in velocity, or change in static pressure. η is the pump efficiency, and may be given by the manufacturer's information, such as in the form of a pump curve, and is typically derived from either fluid dynamics simulation (i.e. solutions to the Navier-stokes for the particular pump geometry), or by testing. The efficiency of the pump will depend upon the pump's configuration and operating conditions (such as rotational speed, fluid density and viscosity etc.)

$$\Delta P = \frac{(v_2^2 - v_1^2)}{2} + \Delta zg + \frac{\Delta p_{\text{static}}}{\rho}$$

For a typical "pumping" configuration, the work is imparted on the fluid, and is thus positive. For the fluid imparting the work on the pump (i.e. a turbine), the work is negative power required to drive the pump is determined by dividing the output power by the pump efficiency. Furthermore, this definition encompasses pumps with no moving parts, such as a siphon.

Pump Efficiency

Pump efficiency is defined as the ratio of the power imparted on the fluid by the pump in relation to the power supplied to drive the pump. Its value is not fixed for a given pump, efficiency is a function of the discharge and therefore also operating head. For centrifugal pumps, the efficiency tends to increase with flow rate up to a point midway through the operating range (peak efficiency) and then declines as flow rates rise further. Pump performance data such as this is usually supplied by the manufacturer before pump selection. Pump efficiencies tend to decline over time due to wear (e.g. increasing clearances as impellers reduce in size).

One important part of system design involves matching the pipeline head loss-flow characteristic with the appropriate pump or pumps in order to operate at or close to the point of maximum efficiency.

Pump efficiency is an important aspect and pumps should be regularly tested. Thermodynamic pump testing is one method.

Pump selection is done by performance curve which is curve between pressure head and flow rate, with power supply also taken care of. Pumps are normally available that run at 50 Hz or 60 Hz.

Pollution of Water

Water pollution is the contamination of water bodies (e.g. lakes, rivers, oceans and groundwater). Water pollution occurs when pollutants are discharged directly or indirectly into water bodies without adequate treatment to remove harmful compounds.

Water pollution affects plants and organisms living in these bodies of water. In almost all cases the effect is damaging not only to individual species and populations, but also to the natural biological communities.

Water pollution occurs when a body of water is adversely affected due to the addition of large amounts of materials to the water. The sources of water pollution are categorized as being a point source or a non-source point of pollution. Point sources of pollution occur when the polluting substance is emitted directly into the waterway. A pipe spewing toxic chemicals directly into a river is an example. A non-point source occurs when there is runoff of pollutants into a waterway, for instance when fertilizer from a field is carried into a stream by surface runoff.

Types of Water Pollution

Toxic Substance —A toxic substance is a chemical pollutant that is not a naturally occurring substance in aquatic ecosystems. The greatest contributors to toxic pollution are herbicides, pesticides and industrial compounds.

Organic Substance — Organic pollution occurs when an excess of organic matter, such as manure or sewage, enters the water. When organic matter increases in a pond, the number of decomposers will increase. These decomposers grow rapidly and use a great deal of oxygen during their growth. This leads to a depletion of oxygen as the decomposition process occurs. A lack of oxygen can kill aquatic organisms. As the aquatic organisms die, they are broken down by decomposers which leads to further depletion of the oxygen levels.

A type of organic pollution can occur when inorganic pollutants such as nitrogen and phosphates accumulate in aquatic ecosystems. High levels of these nutrients cause an overgrowth of plants and algae. As the plants and algae die, they become organic material in the water. The enormous decay of this plant matter, in turn, lowers the oxygen level. The process of rapid plant growth followed by increased activity by decomposers and a depletion of the oxygen level is called *eutrophication.*

Thermal Pollution — Thermal pollution can occur when water is used as a coolant near a power or industrial plant and then is returned to the aquatic environment at a higher temperature than it was originally. Thermal pollution can lead to a decrease in the dissolved oxygen level in the water while also increasing the biological demand of aquatic organisms for oxygen.

Ecological Pollution — Ecological pollution takes place when chemical pollution, organic pollution or thermal pollution are caused by nature rather than by human activity. An example of ecological pollution would be an increased rate of siltation of a waterway after a landslide which would increase the amount of sediments in runoff water. Another example would be when a large animal, such as a deer, drowns in a flood and a large amount of organic material is added to the water as a result. Major geological events such as a volcano eruption might also be sources of ecological pollution.

Specific Sources of Water Pollution

Farming:

- Farms often use large amounts of herbicides and pesticides, both of which are toxic pollutants. These substances are particularly dangerous to life in rivers, streams and lakes, where toxic substances can build up over a period of time.
- Farms also frequently use large amounts of chemical fertilizers that are washed into the waterways and damage the water supply and the life within it. Fertilizers can increase the amounts of nitrates and phosphates in the water, which can lead to the process of eutrophication.
- Allowing livestock to graze near water sources often results in organic waste products being washed into the waterways. This sudden introduction of organic material increaces the amount of nitrogen in the water, and can also lead to eutrophication.
- Four hundred million tons of soil are carried by the Mississippi River to the Gulf of Mexico each year. A great deal of this siltation is due to runoff from the exposed soil of agricultural fields. Excessive amounts of sediment in waterways can block sunlight, preventing aquatic plants from photosynthesizing, and can suffocate fish by clogging their gills.

Business:

- Clearing of land can lead to erosion of soil into the river.
- Waste and sewage generated by industry can get into the water supply, introducing large organic pollutants into the ecosystem.
- Many industrial and power plants use rivers, streams and lakes to despose of waste heat. The resulting hot water can cause thermal pollution. Thermal pollution can have a disasterous effect on life in an aquatic ecosystem as temperature increaces decreace the amount of oxygen in the water, thereby reducing the number of animals that can survive there.
- Water can become contaminated with toxic or radioactive materials from industry, mine sites and abandoned hazardous waste sites.
- Acid precipitation is caused when the burning of fossil fuels emits sulfur dioxide into the atmosphere. The sulfur dioxide reacts with the water in the atmosphere, creating rainfall which contains sulfuric acid. As acid precipitation falls into lakes, streams and ponds it can lower the overall pH of the waterway, killing vital plant life, thereby affecting the whole food chain. It can also leach heavy metals from the soil into the water, killing fish and other aquatic organisms. Because of this, air pollution is potentially one of the most threatening forms of pollution to aquatic ecosystems.

Homes:

- Sewage generated by houses or runoff from septic tanks into nearby waterways, introduce organic pollutants that can cause eutrophication.
- Fertilizers, herbicides and pesticides used for lawn care can runoff and contaminate the waterway. As with agriculteral fertilizers, home fertilizers can lead to the eutrophication of lakes and rivers.
- Improper disposal of hazardous chemicals down the drain itroduce toxic materials into to the ecosystem, contaminating the water supplies in a way that can harm aquatic organisms.
- Leaks of oil and antifreeze from a car on a driveway can be washed off by the rain into nearby waterways, polluting it.

Water pollution is a major global problem which requires ongoing evaluation and revision of water resource policy at all levels (international down to individual aquifers and wells). It has been suggested that it is the leading worldwide cause of deaths and diseases, and that it accounts for the deaths of more than 14,000 people daily. An estimated 700 million Indians have no access to a proper toilet, and 1,000 Indian children die of diarrheal sickness every day. Some 90% of China's cities suffer from some degree of water pollution, and nearly 500 million people lack access to safe drinking water. In addition to the acute problems of water pollution in developing countries, developed countries continue to struggle with pollution problems as well. In the most recent national report on water quality in the United States, 45 percent of assessed stream miles, 47 percent of assessed lake acres, and 32 percent of assessed bay and estuarine square miles were classified as polluted.

Water is typically referred to as polluted when it is impaired by anthropogenic contaminants and either does not support a human use, such as drinking water, and/or undergoes a marked shift in its ability to support its constituent biotic communities, such as fish. Natural phenomena such as volcanoes, algae blooms, storms, and earthquakes also cause major changes in water quality and the ecological status of water.

Categories

Surface water and groundwater have often been studied and managed as separate resources, although they are interrelated. Surface water seeps through the soil and becomes groundwater. Conversely, groundwater can also feed surface water sources. Sources of surface water pollution are generally grouped into two categories based on their origin.

Point Sources

Point source water pollution refers to contaminants that enter a waterway from a single, identifiable source, such as a pipe or ditch. Examples of sources in this category include discharges from a sewage treatment plant, a factory, or a city storm drain. The U.S. Clean Water Act (CWA) defines point source for regulatory enforcement purposes. The CWA definition of point source was amended in 1987 to include municipal storm sewer systems, as well as industrial stormwater, such as from construction sites.

Non–point Sources

Non–point source pollution refers to diffuse contamination that does not originate from a single discrete source. NPS pollution is often the cumulative effect of small amounts of contaminants gathered from a large area. A common example is the leaching out of nitrogen compounds from fertilized agricultural lands. Nutrient runoff in stormwater from "sheet flow" over an agricultural field or a forest are also cited as examples of NPS pollution.

Contaminated storm water washed off of parking lots, roads and highways, called urban runoff, is sometimes included under the category of NPS pollution. However, this runoff is typically channeled into storm drain systems and discharged through pipes to local surface waters, and is a point source. However where such water is not channeled and drains directly to ground it is a non-point source.

Groundwater Pollution

Interactions between groundwater and surface water are complex. Consequently, groundwater pollution, sometimes referred to as groundwater contamination, is not as easily classified as surface water pollution. By its very nature, groundwater aquifers are susceptible to contamination from sources that may not directly affect surface water bodies, and the distinction of point vs. non-point source may be irrelevant. A spill or ongoing releases of chemical or radionuclide contaminants into soil (located away from a surface water body) may not create point source or non-point source pollution, but can contaminate the aquifer below, defined as a toxin plume. The movement of the plume, called a plume front, may be analyzed through a hydrological transport model or groundwater model. Analysis of groundwater contamination may focus on the soil characteristics and site geology, hydrogeology, hydrology, and the nature of the contaminants.

Causes

The specific contaminants leading to pollution in water include a wide spectrum of chemicals, pathogens, and physical or sensory changes such as elevated temperature and discoloration. While many of the chemicals and substances that are regulated may be naturally occurring (calcium, sodium, iron, manganese, etc.) the concentration is often the key in determining what is a natural component of water, and what is a contaminant. High concentrations of naturally-occurring substances can have negative impacts on aquatic flora and fauna.

Oxygen-depleting substances may be natural materials, such as plant matter (e.g. leaves and grass) as well as man-made chemicals. Other natural and anthropogenic substances may cause turbidity (cloudiness) which blocks light and disrupts plant growth, and clogs the gills of some fish species.

Many of the chemical substances are toxic. Pathogens can produce waterborne diseases in either human or animal hosts. Alteration of water's physical chemistry includes acidity (change in pH), electrical conductivity, temperature, and eutrophication. Eutrophication is an increase in the concentration of chemical nutrients in an ecosystem to an extent that increases in the primary productivity of the ecosystem. Depending on the degree of eutrophication, subsequent negative environmental effects such as anoxia (oxygen depletion) and severe reductions in water quality may occur, affecting fish and other animal populations.

Pathogens

Coliform bacteria are a commonly used bacterial indicator of water pollution, although not an actual cause of disease. Other microorganisms sometimes found in surface waters which have caused human health problems include:

- *Burkholderia pseudomallei*
- *Cryptosporidium parvum*
- *Giardia lamblia*
- *Salmonella*
- *Novovirus* and other viruses
- Parasitic worms (helminths).

High levels of pathogens may result from inadequately treated sewage discharges. This can be caused by a sewage plant designed with less than secondary treatment (more typical in less-developed countries). In developed countries, older cities with aging infrastructure may have leaky sewage collection systems (pipes, pumps, valves), which can cause sanitary sewer overflows. Some cities also have combined sewers, which may discharge untreated sewage during rain storms. Pathogen discharges may also be caused by poorly managed livestock operations.

Chemical and Other Contaminants

Contaminants may include organic and inorganic substances.

Organic water pollutants include:

- Detergents
- Disinfection by-products found in chemically disinfected drinking water, such as chloroform

- Food processing waste, which can include oxygen-demanding substances, fats and grease
- Insecticides and herbicides, a huge range of organohalides and other chemical compounds
- Petroleum hydrocarbons, including fuels (gasoline, diesel fuel, jet fuels, and fuel oil) and lubricants (motor oil), and fuel combustion byproducts, from stormwater runoff
- Tree and bush debris from logging operations
- Volatile organic compounds (VOCs), such as industrial solvents, from improper storage.
- Chlorinated solvents, which are dense non-aqueous phase liquids (DNAPLs), may fall to the bottom of reservoirs, since they don't mix well with water and are denser.
    - o Polychlorinated biphenyl (PCBs)
    - o Trichloroethylene
- Perchlorate
- Various chemical compounds found in personal hygiene and cosmetic products

Inorganic water pollutants include:

- Acidity caused by industrial discharges (especially sulfur dioxide from power plants)
- Ammonia from food processing waste
- Chemical waste as industrial by-products
- Fertilizers containing nutrients—nitrates and phosphates—which are found in stormwater runoff from agriculture, as well as commercial and residential use
- Heavy metals from motor vehicles (via urban stormwater runoff) and acid mine drainage
- Silt (sediment) in runoff from construction sites, logging, slash and burn practices or land clearing sites

Macroscopic pollution—large visible items polluting the water—may be termed "floatables" in an urban stormwater context, or marine debris when found on the open seas, and can include such items as:

- Trash or garbage (e.g. paper, plastic, or food waste) discarded by people on the ground, along with accidental or intentional dumping of rubbish, that are washed by rainfall into storm drains and eventually discharged into surface waters

- Nurdles, small ubiquitous waterborne plastic pellets
- Shipwrecks, large derelict ships

Thermal Pollution

Thermal pollution is the degradation of water quality by any process that changes ambient water temperature.

Thermal pollution is the rise or fall in the temperature of a natural body of water caused by human influence. Thermal pollution, unlike chemical pollution, results in a change in the physical properties of water. A common cause of thermal pollution is the use of water as a coolant by power plants and industrial manufacturers. Elevated water temperatures decreases oxygen levels (which can kill fish) and affects ecosystem composition, such as invasion by new thermophilic species. Urban runoff may also elevate temperature in surface waters.

Thermal pollution can also be caused by the release of very cold water from the base of reservoirs into warmer rivers.

A common cause of thermal pollution is the use of water as a coolant by power plants and industrial manufacturers. When water used as a coolant is returned to the natural environment at a higher temperature, the change in temperature decreases oxygen supply, and affects ecosystem composition. Urban runoff–stormwater discharged to surface waters from roads and parking lots–can also be a source of elevated water temperatures.

When a power plant first opens or shuts down for repair or other causes, fish and other organisms adapted to particular temperature range can be killed by the abrupt change in water temperature known as "thermal shock."

Ecological Effects—warm Water

Elevated temperature typically decreases the level of dissolved oxygen in water. This can harm aquatic animals such as fish, amphibians and copepods. Thermal pollution may also increase the metabolic rate of aquatic animals, as enzyme activity, resulting in these organisms consuming more food in a shorter time than if their environment were not changed. An increased metabolic rate may result in fewer resources; the more adapted organisms moving in may have an advantage over organisms that are not used to the warmer temperature. As a result, food chains of the old and new environments may be compromised. Some fish species will avoid stream segments or coastal areas adjacent to a thermal discharge. Biodiversity can be decreased as a result. High temperature limits oxygen dispersion into deeper waters, contributing to anaerobic conditions. This can lead to increased bacteria levels when there is ample food supply. Many aquatic species will fail to reproduce at elevated temperatures.

Primary producers are affected by warm water because higher water temperature increases plant growth rates, resulting in a shorter lifespan and species overpopulation. This can cause an algae bloom which reduces oxygen levels.

Temperature changes of even one to two degrees Celsius can cause significant changes in organism metabolism and other adverse cellular biology effects. Principal adverse changes can include rendering cell walls less permeable to necessary osmosis, coagulation of cell proteins, and alteration of enzyme metabolism. These cellular level effects can adversely affect mortality and reproduction.

A large increase in temperature can lead to the denaturing of life-supporting enzymes by breaking down hydrogen- and disulphide bonds within the quaternary structure of the enzymes. Decreased enzyme activity in aquatic organisms can cause problems such as the inability to break down lipids, which leads to malnutrition.

In limited cases, warm water has little deleterious effect and may even lead to improved function of the receiving aquatic ecosystem. This phenomenon is seen especially in seasonal waters and is known as thermal enrichment. An extreme case is derived from the aggregational habits of the manatee, which often uses power plant discharge sites during winter. Projections suggest that manatee populations would decline upon the removal of these discharges.

Ecological Effects — Cold Water

Releases of unnaturally cold water from reservoirs can dramatically change the fish and macroinvertebrate fauna of rivers, and reduce river productivity. In Australia, where many rivers have warmer temperature regimes, native fish species have been eliminated, and macroinvertebrate fauna have been drastically altered. Due to the sudden fall of water temperature the contraction on dam and bridge pylon may take place.

12

Water and Habitable Zone

The existence of liquid water, and to a lesser extent its gaseous and solid forms, on Earth are vital to the existence of life on Earth as we know it. The Earth is located in the habitable zone of the solar system; if it were slightly closer to or farther from the Sun (about 5%, or about 8 million kilometers), the conditions which allow the three forms to be present simultaneously would be far less likely to exist.

Earth's gravity allows it to hold an atmosphere. Water vapor and carbon dioxide in the atmosphere provide a temperature buffer (greenhouse effect) which helps maintain a relatively steady surface temperature. If Earth were smaller, a thinner atmosphere would allow temperature extremes, thus preventing the accumulation of water except in polar ice caps (as on Mars).

The surface temperature of Earth has been relatively constant through geologic time despite varying levels of incoming solar radiation (insolation), indicating that a dynamic process governs Earth's temperature via a combination of greenhouse gases and surface or atmospheric albedo. This proposal is known as the *Gaia hypothesis*.

The state of water on a planet depends on ambient pressure, which is determined by the planet's gravity. If a planet is sufficiently massive, the water on it may be solid even at high temperatures, because of the high pressure caused by gravity, as it was observed on exoplanets Gliese 436 b and GJ 1214 b.

There are various theories about origin of water on Earth.

On Earth

Hydrology is the study of the movement, distribution, and quality of water throughout the Earth. The study of the distribution of water is hydrography. The study of the distribution and movement of groundwater is hydrogeology, of glaciers is glaciology, of inland waters is limnology and distribution of oceans is oceanography. Ecological processes with hydrology are in focus of ecohydrology.

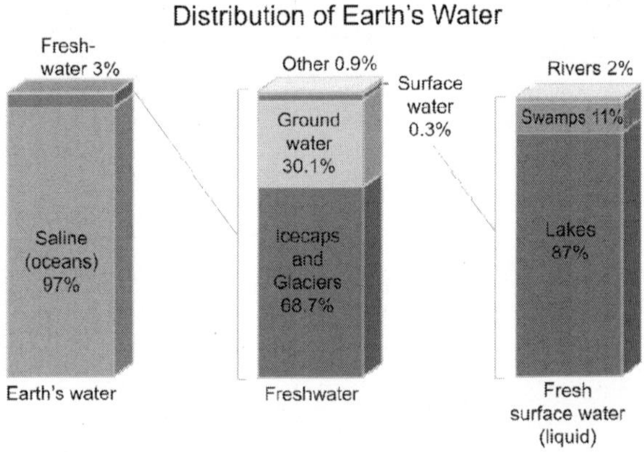

Figure: *A graphical distribution of the locations of water on Earth.*

The collective mass of water found on, under, and over the surface of a planet is called the hydrosphere. Earth's approximate water volume (the total water supply of the world) is 1,338,000,000 km^3 (321,000,000 mi^3).

Liquid water is found in bodies of water, such as an ocean, sea, lake, river, stream, canal, pond, or puddle. The majority of water on Earth is sea water. Water is also present in the atmosphere in solid, liquid, and vapor states. It also exists as groundwater in aquifers.

Water is important in many geological processes. Groundwater is present in most rocks, and the pressure of this groundwater affects patterns of faulting. Water in the mantle is responsible for the melt that produces volcanoes at subduction zones. On the surface of the Earth, water is important in both chemical and physical weathering processes. Water and, to a lesser but still significant extent, ice, are also responsible for a large amount of sediment transport that occurs on the surface of the earth. Deposition of transported sediment forms many types of sedimentary rocks, which make up the geologic record of Earth history.

Water Cycle

The water cycle (known scientifically as the hydrologic cycle) refers to the continuous exchange of water within the hydrosphere, between the atmosphere, soil water, surface water, groundwater, and plants.

Water moves perpetually through each of these regions in the *water cycle* consisting of following transfer processes:

- evaporation from oceans and other water bodies into the air and transpiration from land plants and animals into air.
- precipitation, from water vapor condensing from the air and falling to earth or ocean.
- runoff from the land usually reaching the sea.

Most water vapor over the oceans returns to the oceans, but winds carry water vapor over land at the same rate as runoff into the sea, about 47 Tt per year. Over land, evaporation and transpiration contribute another 72 Tt per year. Precipitation, at a rate of 119 Tt per year over land, has several forms: most commonly rain, snow, and hail, with some contribution from fog and dew. Condensed water in the air may also refract sunlight to produce rainbows.

Water runoff often collects over watersheds flowing into rivers. A mathematical model used to simulate river or stream flow and calculate water quality parameters is hydrological transport model. Some of water is diverted to irrigation for agriculture. Rivers and seas offer opportunity for travel and commerce. Through erosion, runoff shapes the environment creating river valleys and deltas which provide rich soil and level ground for the establishment of population centers. A flood occurs when an area of land, usually low-lying, is covered with water. It is when a river overflows its banks or flood from the sea. A drought is an extended period of months or years when a region notes a deficiency in its water supply. This occurs when a region receives consistently below average precipitation.

Water Resources

Water resources are sources of water that are useful or potentially useful. Uses of water include agricultural, industrial, household, recreational and environmental activities. Virtually all of these human uses require fresh water.

Ninety-seven percent of the water on the Earth is salt water. Only three percent is fresh water; slightly over two thirds of this is frozen in glaciers and polar ice caps. The remaining unfrozen freshwater is found mainly as groundwater, with only a small fraction present above ground or in the air.

Fresh water is a renewable resource, yet the world's supply of clean, fresh water is steadily decreasing. Water demand already exceeds supply in many parts of the world and as the world population continues to rise, so too does the water demand. Awareness of the global importance of preserving water for ecosystem services has only recently emerged as, during the 20th century, more than half the world's wetlands have been lost along with their valuable environmental services for Water Education. The framework for allocating water resources to water users (where such a framework exists) is known as water rights.

Sources of Fresh Water

Surface Water

Surface water is water in a river, lake or fresh water wetland. Surface water is naturally replenished by precipitation and naturally lost through discharge to the oceans, evaporation, evapotranspiration and sub-surface seepage.

Although the only natural input to any surface water system is precipitation within its watershed, the total quantity of water in that system at any given time is also dependent on many other factors. These factors include storage capacity in lakes, wetlands and artificial reservoirs, the permeability of the soil beneath these storage bodies, the runoff characteristics of the land in the watershed, the timing of the precipitation and local evaporation rates. All of these factors also affect the proportions of water lost.

Human activities can have a large and sometimes devastating impact on these factors. Humans often increase storage capacity by constructing reservoirs and decrease it by draining wetlands. Humans often increase runoff quantities and velocities by paving areas and channelizing stream flow.

The total quantity of water available at any given time is an important consideration. Some human water users have an intermittent need for water. For example, many farms require large quantities of water in the spring, and no water at all in the winter. To supply such a farm with water, a surface water system may require a large storage capacity to collect water throughout the year and release it in a short period of time. Other users have a continuous need for water, such as a power plant that requires water for cooling. To supply such a power plant with water, a surface water system only needs enough storage capacity to fill in when average stream flow is below the power plant's need.

Nevertheless, over the long term the average rate of precipitation within a watershed is the upper bound for average consumption of natural surface water from that watershed.

Natural surface water can be augmented by importing surface water from another watershed through a canal or pipeline. It can also be artificially augmented from any of the other sources listed here, however in practice the quantities are negligible. Humans can also cause surface water to be "lost" (i.e. become unusable) through pollution.

Brazil is the country estimated to have the largest supply of fresh water in the world, followed by Russia and Canada.

Under River Flow

Throughout the course of a river, the total volume of water transported downstream will often be a combination of the visible free water flow together with a substantial contribution flowing through sub-surface rocks and gravels that underlie the river and its floodplain called the hyporheic zone. For many rivers in large valleys, this unseen component of flow may greatly exceed the visible flow. The hyporheic zone often forms a dynamic interface between surface water and true ground-water receiving water from the ground water when aquifers are fully charged and contributing water to ground-water when ground waters are depleted. This is especially significant in karst areas where pot-holes and underground rivers are common.

Ground Water

Groundwater is water located beneath the ground surface in soil pore spaces and in the fractures of rock formations. A unit of rock or an unconsolidated deposit is called an aquifer when it can yield a usable quantity of water. The depth at which soil pore spaces or fractures and voids in rock become completely saturated with water is called the water table. Groundwater is recharged from, and eventually flows to, the surface naturally; natural discharge often occurs at springs and seeps, and can form oases or wetlands. Groundwater is also often withdrawn for agricultural, municipal and industrial use by constructing and operating extraction wells. The study of the distribution and movement of groundwater is hydrogeology, also called groundwater hydrology.

Typically, groundwater is thought of as liquid water flowing through shallow aquifers, but technically it can also include soil moisture, permafrost (frozen soil), immobile water in very low permeability bedrock, and deep geothermal or oil formation water. Groundwater is hypothesized to provide lubrication that can possibly influence the movement of faults. It is likely that much of the Earth's subsurface contains some water, which may be mixed with other fluids in some instances. Groundwater may not be confined only to the Earth. The formation of some of the landforms observed on Mars may have been influenced by groundwater. There is also evidence that liquid water may also exist in the subsurface of Jupiter's moon Europa.

Sub-surface water, or groundwater, is fresh water located in the pore space of soil and rocks. It is also water that is flowing within aquifers below the water table. Sometimes it is useful to make a distinction between sub-surface water that is closely associated with surface water and deep sub-surface water in an aquifer (sometimes called "fossil water").

Sub-surface water can be thought of in the same terms as surface water: inputs, outputs and storage. The critical difference is that due to its slow rate of turnover, sub-surface water storage is generally much larger compared to inputs than it is for surface water. This difference makes it easy for humans to use sub-surface water unsustainably for a long time without severe consequences. Nevertheless, over the long term the average rate of seepage above a sub-surface water source is the upper bound for average consumption of water from that source.

The natural input to sub-surface water is seepage from surface water. The natural outputs from sub-surface water are springs and seepage to the oceans.

If the surface water source is also subject to substantial evaporation, a sub-surface water source may become saline. This situation can occur naturally under endorheic bodies of water, or artificially under irrigated farmland. In coastal areas, human use of a sub-surface water source may cause the direction of seepage to ocean to reverse which can also cause soil salinization. Humans can also cause sub-surface water to be "lost" (i.e. become unusable) through pollution. Humans can increase the input to a sub-surface water source by building reservoirs or detention ponds.

Desalination

Desalination is an artificial process by which saline water (generally sea water) is converted to fresh water. The most common desalination processes are distillation and reverse osmosis. Desalination is currently expensive compared to most alternative sources of water, and only a very small fraction of total human use is satisfied by desalination. It is only economically practical for high-valued uses (such as household and industrial uses) in arid areas. The most extensive use is in the Persian Gulf.

Frozen Water

Several schemes have been proposed to make use of icebergs as a water source, however to date this has only been done for novelty purposes. Glacier runoff is considered to be surface water.

The Himalayas, which are often called "The Roof of the World", contain some of the most extensive and rough high altitude areas on Earth as well as the greatest area of glaciers and permafrost outside of the poles. Ten of Asia's largest rivers flow from there, and more than a billion people's livelihoods depend on them. To complicate matters, temperatures are rising more rapidly here than the global average. In Nepal the temperature has risen with 0.6 degree over the last decade, whereas the global warming has been around 0.7 over the last hundred years.

Uses of Fresh Water

Uses of fresh water can be categorized as consumptive and non-consumptive (sometimes called "renewable"). A use of water is consumptive if that water is not immediately available for another use. Losses to sub-surface seepage and evaporation are considered consumptive, as is water incorporated into a product (such as farm produce). Water that can be treated and returned as surface water, such as sewage, is generally considered non-consumptive if that water can be put to additional use. Water use in power generation and industry is generally described using an alternate terminology, focusing on separate measurements of withdrawal and consumption. Withdrawal describes the removal of water from the environment, while consumption describes the conversion of fresh water into some other form, such as atmospheric water vapor or contaminated waste water

Agricultural

It is estimated that 69% of worldwide water use is for irrigation, with 15-35% of irrigation withdrawals being unsustainable. It takes around 3,000 litres of water, converted from liquid to vapour, to produce enough food to satisfy one person's daily dietary need. This is a considerable amount, when compared to that required for drinking, which is between two and five litres. To produce food for the now over 7 billion people who inhabit the planet today requires the water that would fill a canal ten metres deep, 100 metres wide and 7.1 million kilometres long – that's enough to circle the globe 180 times.

Increasing Water Scarcity

Fifty years ago, the common perception was that water was an infinite resource. At this time, there were fewer than half the current number of people on the planet. People were not as wealthy as today, consumed fewer calories and ate less meat, so less water was needed to produce their food. They required a third of the volume of water we presently take from rivers. Today, the competition for water resources is much more intense. This is because there are now nearly seven billion people on the planet, their consumption of water-thirsty meat and vegetables is rising, and there is increasing competition for water from industry, urbanisation and biofuel crops. In future, even more water will be needed to produce food because the Earth's population is forecast to rise to 9 billion by 2050. An additional 2.5 or 3 billion people, choosing to eat fewer cereals and more meat and vegetables could add an additional five million kilometres to the virtual canal mentioned above.

An assessment of water management in agriculture was conducted in 2007 by the International Water Management Institute in Sri Lanka to see if the world had sufficient water to provide food for its growing population. It assessed the current availability of water for agriculture on a global scale and mapped out locations suffering from water scarcity. It found that a fifth of the world's people, more than 1.2 billion, live in areas of physical water scarcity, where there is not enough water to meet all demands. A further 1.6 billion people live in areas experiencing economic water scarcity, where the lack of investment in water or insufficient human capacity make it impossible for authorities to satisfy the demand for water. The report found that it would be possible to produce the food required in future, but that continuation of today's food production and environmental trends would lead to crises in many parts of the world. To avoid a global water crisis, farmers will have to strive to increase productivity to meet growing demands for food, while industry and cities find ways to use water more efficiently.

In some areas of the world irrigation is necessary to grow any crop at all, in other areas it permits more profitable crops to be grown or enhances crop yield. Various irrigation methods involve different trade-offs between crop yield, water consumption and capital cost of equipment and structures. Irrigation methods such as furrow and overhead sprinkler irrigation are usually less expensive but are also typically less efficient, because much of the water evaporates, runs off or drains below the root zone. Other irrigation methods considered to be more efficient include drip or trickle irrigation, surge irrigation, and some types of sprinkler systems where the sprinklers are operated near ground level. These types of systems, while more expensive, usually offer greater potential to minimize runoff, drainage and evaporation. Any system that is improperly managed can be wasteful, all methods have the potential for high efficiencies under suitable conditions, appropriate irrigation timing and management. Some issues that are often insufficiently considered are salinization of sub-surface water and contaminant accumulation leading to water quality declines.

As global populations grow, and as demand for food increases in a world with a fixed water supply, there are efforts under way to learn how to produce more food with less water, through improvements in irrigation methods and technologies, agricultural water management, crop types, and water monitoring. Aquaculture is a small but growing agricultural use of water. Freshwater commercial fisheries may also be considered as agricultural uses of water, but have generally been assigned a lower priority than irrigation.

Industrial

It is estimated that 22% of worldwide water use is industrial. Major industrial users include hydroelectric dams, thermoelectric power plants, which use water for cooling, ore and oil refineries, which use water in chemical processes, and manufacturing plants, which use water as a solvent. Water withdrawal can be very high for certain industries, but consumption is generally much lower than that of agriculture. Water is used in renewable power generation. Hydroelectric power derives energy from the force of water flowing downhill, driving a turbine connected to a generator. This hydroelectricity is a low-cost, non-polluting, renewable energy source. Significantly, hydroelectric power can also be used for load following unlike most renewable energy sources which are intermittent. Ultimately, the energy in a hydroelectric powerplant is supplied by the sun. Heat from the sun evaporates water, which condenses as rain in higher altitudes and flows downhill. Pumped-storage hydroelectric plants also exist, which use grid electricity to pump water uphill when demand is low, and use to stored water to produce electricity when demand is high.

Hydroelectric power plants generally require the creation of a large artificial lake. Evaporation from this lake is higher than evaporation from a river due to the larger surface area exposed to the elements, resulting in much higher water consumption. The process of driving water through the turbine and tunnels or pipes also briefly removes this water from the natural environment, creating water withdrawal. The impact of this withdrawal on wildlife varies greatly depending on the design of the powerplant.

Pressurized water is used in water blasting and water jet cutters. Also, very high pressure water guns are used for precise cutting. It works very well, is relatively safe, and is not harmful to the environment. It is also used in the cooling of machinery to prevent overheating, or prevent saw blades from overheating. This is generally a very small source of water consumption relative to other uses.

Water is also used in many large scale industrial processes, such as thermoelectric power production, oil refining, fertilizer production and other chemical plant use, and natural gas extraction from shale rock. Discharge of untreated water from industrial uses is pollution. Pollution includes discharged solutes (chemical pollution) and increased water temperature (thermal pollution). Industry requires pure water for many applications and utilizes a variety of purification techniques both in water supply and discharge. Most of this pure water is generated on site, either from natural freshwater or from municipal grey water. Industrial consumption of water is generally much lower than withdrawal, due to laws requiring industrial grey water to be treated and returned to the environment. Thermoelectric powerplants using cooling towers have high consumption, nearly equal to their withdrawal, as most of the withdrawn water is evaporated as part of the cooling process. The withdrawal, however, is lower than in once-through cooling systems.

Household

It is estimated that 8% of worldwide water use is for household purposes. These include drinking water, bathing, cooking, sanitation, and gardening. Basic household water requirements have been estimated by Peter Gleick at around 50 liters per person per day, excluding water for gardens. Drinking water is water that is of sufficiently high quality so that it can be consumed or used without risk of immediate or long term harm. Such water is commonly called potable water. In most developed countries, the water supplied to households, commerce and industry is all of drinking water standard even though only a very small proportion is actually consumed or used in food preparation.

Fresh Water Storage

Some runoff water is trapped for periods of time, for example in lakes. At high altitude, during winter, and in the far north and south, snow collects in ice caps, snow pack and glaciers. Water also infiltrates the ground and goes into aquifers. This groundwater later flows back to the surface in springs, or more spectacularly in hot springs and geysers. Groundwater is also extracted artificially in wells. This water storage is important, since clean, fresh water is essential to human and other land-based life. In many parts of the world, it is in short supply.

Sea Water

Sea water contains about 3.5% salt on average, plus smaller amounts of other substances. The physical properties of sea water differ from fresh water in some important respects. It freezes at a lower temperature (about "1.9 °C) and its density increases with decreasing temperature to the freezing point, instead of reaching maximum density at a temperature above freezing. The salinity of water in major seas varies from about 0.7% in the Baltic Sea to 4.0% in the Red Sea.

Tides

Tides are the cyclic rising and falling of local sea levels caused by the tidal forces of the Moon and the Sun acting on the oceans. Tides cause changes in the depth of the marine and estuarine water bodies and produce oscillating currents known as tidal streams. The changing tide produced at a given location is the result of the changing positions of the Moon and Sun relative to the Earth coupled with the effects of Earth rotation and the local bathymetry. The strip of seashore that is submerged at high tide and exposed at low tide, the intertidal zone, is an important ecological product of ocean tides.

Effects on Life

From a biological standpoint, water has many distinct properties that are critical for the proliferation of life that set it apart from other substances. It carries out this role by allowing organic compounds to react in ways that ultimately allow replication. All known forms of life depend on water. Water is vital both as a solvent in which many of the body's solutes dissolve and as an essential part of many metabolic processes within the body. Metabolism is the sum total of anabolism and catabolism. In anabolism, water is removed from molecules (through energy requiring enzymatic chemical reactions) in order to grow larger molecules (e.g. starches, triglycerides and proteins for storage of fuels and information). In catabolism, water is used to break bonds in order to generate smaller molecules (e.g. glucose, fatty acids and amino acids to be used for fuels for energy use or other purposes). Without water, these particular metabolic processes could not exist.

Water is fundamental to photosynthesis and respiration. Photosynthetic cells use the sun's energy to split off water's hydrogen from oxygen. Hydrogen is combined with CO_2 (absorbed from air or water) to form glucose and release oxygen. All living cells use such fuels and oxidize the hydrogen and carbon to capture the sun's energy and reform water and CO_2 in the process (cellular respiration).

Water is also central to acid-base neutrality and enzyme function. An acid, a hydrogen ion (H^+, that is, a proton) donor, can be neutralized by a base, a proton acceptor such as hydroxide ion (O≈) to form water. Water is considered to be neutral, with a pH (the negative log of the hydrogen ion concentration) of 7. Acids have pH values less than 7 while bases have values greater than 7.

Aquatic Life Forms

Earth surface waters are filled with life. The earliest life forms appeared in water; nearly all fish live exclusively in water, and there are many types of marine mammals, such as dolphins and whales.

Some kinds of animals, such as amphibians, spend portions of their lives in water and portions on land. Plants such as kelp and algae grow in the water and are the basis for some underwater ecosystems. Plankton is generally the foundation of the ocean food chain.

Aquatic vertebrates must obtain oxygen to survive, and they do so in various ways. Fish have gills instead of lungs, although some species of fish, such as the lungfish, have both. Marine mammals, such as dolphins, whales, otters, and seals need to surface periodically to breathe air. Some amphibians are able to absorb oxygen through their skin. Invertebrates exhibit a wide range of modifications to survive in poorly oxygenated waters including breathing tubes and gills (*Carcinus*). However as invertebrate life evolved in an aquatic habitat most have little or no specialisation for respiration in water.

Effects on Human Civilization

Civilization has historically flourished around rivers and major waterways; Mesopotamia, the so-called cradle of civilization, was situated between the major rivers Tigris and Euphrates; the ancient society of the Egyptians depended entirely upon the Nile. Large metropolises like Rotterdam, London, Montreal, Paris, New York City, Buenos Aires, Shanghai, Tokyo, Chicago, and Hong Kong owe their success in part to their easy accessibility via water and the resultant expansion of trade. Islands with safe water ports, like Singapore, have flourished for the same reason. In places such as North Africa and the Middle East, where water is more scarce, access to clean drinking water was and is a major factor in human development.

Health and Pollution

Water fit for human consumption is called drinking water or potable water. Water that is not potable may be made potable by filtration or distillation, or by a range of other methods.

Water that is not fit for drinking but is not harmful for humans when used for swimming or bathing is called by various names other than potable or drinking water, and is sometimes called safe water, or "safe for bathing". Chlorine is a skin and mucous membrane irritant that is used to make water safe for bathing or drinking. Its use is highly technical and is usually monitored by government regulations (typically 1 part per million (ppm) for drinking water, and 1–2 ppm of chlorine not yet reacted with impurities for bathing water). Water for bathing may be maintained in satisfactory microbiological condition using chemical disinfectants such as chlorine or ozone or by the use of ultraviolet light.

In the USA, non-potable forms of wastewater generated by humans may be referred to as greywater, which is treatable and thus easily able to be made potable again, and blackwater, which generally contains sewage and other forms of waste which require further treatment in order to be made reusable. Greywater composes 50–80% of residential wastewater generated by a household's sanitation equipment (sinks, showers and kitchen runoff, but not toilets, which generate blackwater.) These terms may have different meanings in other countries and cultures.

This natural resource is becoming scarcer in certain places, and its availability is a major social and economic concern. Currently, about a billion people around the world routinely drink unhealthy water. Most countries accepted the goal of halving by 2015 the number of people worldwide who do not have access to safe water and sanitation during the 2003 G8 Evian summit. Even if this difficult goal is met, it will still leave more than an estimated half a billion people without access to safe drinking water and over a billion without access to adequate sanitation. Poor water quality and bad sanitation are deadly; some five million deaths a year are caused by polluted drinking water. The World Health Organization estimates that safe water could prevent 1.4 million child deaths from diarrhea each year. Water, however, is not a finite resource, but rather re-circulated as potable water in precipitation in quantities many degrees of magnitude higher than human consumption. Therefore, it is the relatively small quantity of water in reserve in the earth (about 1% of our drinking water supply, which is replenished in aquifers around every 1 to 10 years), that is a non-renewable resource, and it is, rather, the distribution of potable and irrigation water which is scarce, rather than the actual amount of it that exists on the earth. Water-poor countries use importation of goods as the primary method of importing water (to leave enough for local human consumption), since the manufacturing process uses around 10 to 100 times products' masses in water.

In the developing world, 90% of all wastewater still goes untreated into local rivers and streams. Some 50 countries, with roughly a third of the world's population, also suffer from medium or high water stress, and 17 of these extract more water annually than is recharged through their natural water cycles. The strain not only affects surface freshwater bodies like rivers and lakes, but it also degrades groundwater resources.

Human Uses

Water Supply: Water supply is the provision of water by public utilities, commercial organisations, community endeavours or by individuals, usually via a system of pumps and pipes. Irrigation is covered separately.

Global Access to Clean Water

In 2010 about 84% of the global population (6.74 billion people) had access to piped water supply through house connections or to an improved water source through other means than house, including standpipes, "water kiosks", protected springs and protected wells. However, about 14% (884 million people) did not have access to an improved water source and had to use unprotected wells or springs, canals, lakes or rivers for their water needs.

A clean water supply, especially so with regard to sewage, is the single most important determinant of public health. Destruction of water supply and/or sewage disposal infrastructure after major catastrophes (earthquakes, floods, war, etc.) poses the immediate threat of severe epidemics of waterborne diseases, several of which can be life-threatening.

Technical Overview

Water supply systems get water from a variety of locations, including groundwater (aquifers), surface water (lakes and rivers), conservation and the sea through desalination. The water is then, in most cases, purified, disinfected through chlorination and sometimes fluoridated. Treated water then either flows by gravity or is pumped to reservoirs, which can be elevated such as water towers or on the ground (for indicators related to the efficiency of drinking water distribution see non-revenue water). Once water is used, wastewater is typically discharged in a sewer system and treated in a wastewater treatment plant before being discharged into a river, lake or the sea or reused for landscaping, irrigation or industrial use.

Water Supply Network

A water supply system or water supply network is a system of engineered hydrologic and hydraulic components which provide water supply. A water supply system typically includes:

1. A drainage basin;
2. A raw (untreated) water collection point (above or below ground) where the water accumulates, such as a lake, a river, or groundwater from an underground aquifer. Untreated drinking water (usually water being transferred to the water purification facilities) may be transferred using uncovered ground-level aqueducts, covered tunnels or underground water pipes.
3. Water purification facilities. Treated water is transferred using water pipes (usually underground).
4. Water storage facilities such as reservoirs, water tanks, or watertowers. Smaller water systems may store the water in cisterns or pressure vessels. (Tall buildings may also need to store water locally in pressure vessels in order for the water to reach the upper floors.)
5. Additional water pressurizing components such as pumping stations may need to be situated at the outlet of underground or above ground reservoirs or cisterns (if gravity flow is impractical)
6. A pipe network for distribution of water to the consumers (which may be private houses or industrial, commercial or institution establishments) and other usage points (such as fire hydrants)
7. Connections to the sewers (underground pipes, or aboveground ditches in some developing countries) are generally found downstream of the water consumers, but the sewer system is considered to be a separate system, rather than part of the water supply system

Water Abstraction and Raw Water Transfer

Water abstraction, water extraction, or groundwater abstraction is the process of taking water from any source, either temporarily or permanently. Most water is used for irrigation or treatment to produce drinking water.

Depending on the environmental legislation in the relevant country, controls may be placed on abstraction to limit the amount of water that can be removed. Over abstraction can lead to rivers drying up or the level of groundwater aquifers reducing unacceptably.

The science of hydrogeology is used to assess safe abstraction levels. In some U.S. industries, the terminology for water use describes the removal of water from the environment as 'water withdrawal', and the conversion of fresh water into water vapor or wastewater as 'water consumption'. Raw water (untreated) is collected from a surface water source (such as an intake on a lake or a river) or from a groundwater source (such as a water well drawing from an underground aquifer) within the watershed that provides the water resource.

Shallow dams and reservoirs are susceptible to outbreaks of toxic algae, especially if the water is warmed by a hot sun. The bacteria grow from stormwater runoff carrying fertilizer into the river where it acts as a nutrient for the algae. Such outbreaks render the water unfit for human consumption. The raw water is transferred to the water purification facilities using uncovered aqueducts, covered tunnels or underground water pipes.

Water Treatment

Virtually all large systems must treat the water; a fact that is tightly regulated by global, state and federal agencies, such as the World Health Organization (WHO) or the United States Environmental Protection Agency (EPA). Water treatment must occur before the product reaches the consumer and afterwards (when it is discharged again). Water purification usually occurs close to the final delivery points to reduce pumping costs and the chances of the water becoming contaminated after treatment.

Traditional surface water treatment plants generally consists of three steps: clarification, filtration and disinfection. Clarification refers to the separation of particles (dirt, organic matter, etc.) from the water stream. Chemical addition (i.e. alum, ferric chloride) destabilizes the particle charges and prepares them for clarification either by settling or floating out of the water stream. Sand, anthracite or activated carbon filters refine the water stream, removing smaller particulate matter. While other methods of disinfection exist, the preferred method is via chlorine addition. Chlorine effectively kills bacteria and most viruses and maintains a residual to protect the water supply through the supply network.

Water Distribution Network

The product, delivered to the point of consumption, is called fresh water if it receives little or no treatment, or drinking water if the treatment achieves the water quality standards required for human consumption. Once treated, chlorine is added to the water and it is distributed by the local supply network. Today, water supply systems are typically constructed of plastic, ferrous, or concrete circular pipe.

However, other "pipe" shapes and material may be used, such as square or rectangular concrete boxes, arched brick pipe, or wood. Near the end point, the network of pipes through which the water is delivered is often referred to as the *water mains*.

The energy that the system needs to deliver the water is called pressure. That energy is transferred to the water, therefore becoming water pressure, in a number of ways: by a pump, by gravity feed from a water source (such as a water tower) at a higher elevation, or by compressed air. The water is often transferred from a water reserve such as a large communal reservoir before being transported to a more pressurised reserve such as a watertower. In small domestic systems, the water may be pressurised by a pressure vessel or even by an underground cistern (the latter however does need additional pressurizing). This eliminates the need of a water-tower or any other heightened water reserve to supply the water pressure.

These systems are usually owned and maintained by local governments, such as cities, or other public entities, but are occasionally operated by a commercial enterprise. Water supply networks are part of the master planning of communities, counties, and municipalities. Their planning and design requires the expertise of city planners and civil engineers, who must consider many factors, such as location, current demand, future growth, leakage, pressure, pipe size, pressure loss, fire fighting flows, etc. — using pipe network analysis and other tools. Constructioncomparable sewage systems, was one of the great engineering advances that made urbanization possible. Improvement in the quality of the water has been one of the great advances in public health. As water passes through the distribution system, the water quality can degrade by chemical reactions and biological processes. Corrosion of metal pipe materials in the distribution system can cause the release of metals into the water with undesirable aesthetic and health effects. Release of iron from unlined iron pipes can result in customer reports of "red water" at the tap . Release of copper from copper pipes can result in customer reports of "blue water" and/or a metallic taste. Release of lead can occur from the solder used to join copper pipe together or from brass fixtures. Copper and lead levels at the consumer's tap are regulated to protect consumer health.

Utilities will often adjust the chemistry of the water before distribution to minimize its corrosiveness. The simplest adjustment involves control of pH and alkalinity to produce a water that tends to passivate corrosion by depositing a layer of calcium carbonate. Corrosion inhibitors are often added to reduce release of metals into the water. Common corrosion inhibitors added to the water are phosphates and silicates. Maintenance of a biologically safe drinking water is another goal in water distribution. Typically, a chlorine based disinfectant, such as sodium hypochlorite or monochloramine is added to the water as it leaves the treatment plant. Booster stations can be placed within the distribution system to ensure that all areas of the distribution system have adequate sustained levels of disinfection.

Topologies of Water Distribution Networks

Like electric power lines, roads, and microwave radio networks, water systems may have a loop or branch network topology, or a combination of both. The piping networks are circular or rectangular. If any one section of water distribution main fails or needs repair, that section can be isolated without disrupting all users on the network.

Most systems are divided into zones. Factors determining the extent or size of a zone can include hydraulics, telemetry systems, history, and population density. Sometimes systems are designed for a specific area then are modified to accommodate development. Terrain affects hydraulics and some forms of telemetry. While each zone may operate as a stand-alone system, there is usually some arrangement to interconnect zones in order to manage equipment failures or system failures.

Water Network Maintenance

Water supply networks usually represent the majority of assets of a water utility. Systematic documentation of maintenance works using a Computerized Maintenance Management System is a key to a successful operation of a water utility.

Service Quality

Many of the 3.5 billion people having access to piped water receive a poor or very poor quality of service, especially in developing countries where about 80% of the world population lives. Water supply service quality has many dimensions: continuity; water quality; pressure; and the degree of responsiveness of service providers to customer complaints.

Continuity of Supply

Continuity of water supply is taken for granted in most developed countries, but is a severe problem in many developing countries, where sometimes water is only provided for a few hours every day or a few days a week. It is estimated that about half of the population of developing countries receives water on an intermittent basis.

Water Quality

Drinking water quality has a micro-biological and a physico-chemical dimension. There are thousands of parameters of water quality. In public water supply systems water should, at a minimum, be disinfected — most commonly through the use of chlorination or the use of ultra violet light — or it may need to undergo treatment, especially in the case of surface water.

Water Pressure

Water pressures vary in different locations of a distribution system. Water mains below the street may operate at higher pressures, with a pressure reducer located at each point where the water enters a building or a house. In poorly managed systems, water pressure can be so low as to result only in a trickle of water or so high that it leads to damage to plumbing fixtures and waste of water. Pressure in an urban water system is typically maintained either by a pressurized water tank serving an urban area, by pumping the water up into a tower and relying on gravity to maintain a constant pressure in the system or solely by pumps at the water treatment plant and repeater pumping stations.

Typical UK pressures are 4-5 bar for an urban supply. However, some people can get over eight bars or below one bar. A single iron main pipe may cross a deep valley, it will have the same nominal pressure, however each consumer will get a bit more or less because of the hydrostatic pressure (about 1 bar/10m height). So people at the bottom of a 100-foot (30 m) hill will get about 3 bars more than those at the top.

The effective pressure also varies because of the supply resistance even for the same static pressure. An urban consumer may have 5 metres of ½" lead pipe running from the iron main, so the kitchen tap flow will be fairly unrestricted, so high flow. A rural consumer may have a kilometre of rusted and limed ¾" iron pipe so their kitchen tap flow will be small.

For this reason the UK domestic water system has traditionally (prior to 1989) employed a "cistern feed" system, where the incoming supply is connected to the kitchen sink and also a header/storage tank in the attic. Water can dribble into this tank through a ½" lead pipe, plus ball valve, and then supply the house on 22 or 28 mm pipes. Gravity water has a small pressure (say ¼ bar in the bathroom) but needs wide pipes allow higher flows. This is fine for baths and toilets but is frequently inadequate for showers. People install shower booster pumps to increase the pressure. For this reason urban houses are increasingly using mains pressure boilers (combies) which take a long time to fill a bath but suit the high back pressure of a shower.

Comparing the Performance of Water and Sanitation Service Providers

Comparing the performance of water and sanitation service providers (utilities) is needed, because the sector offers limited scope for direct competition (natural monopoly). Firms operating in competitive markets are under constant pressure to out perform each other. Water utilities are often sheltered from this pressure, and it frequently shows: some utilities are on a sustained improvement track, but many others keep falling further behind best practice. Benchmarking the performance of utilities allows to simulate competition, establish realistic targets for improvement and create pressure to catch up with better utilities. Information on benchmarks for water and sanitation utilities is provided by the International Benchmarking Network for Water and Sanitation Utilities.

Institutional Responsibility and Governance

A great variety of institutions have responsibilities in water supply. A basic distinction is between institutions responsible for policy and regulation on the one hand; and institutions in charge of providing services on the other hand.

Policy and Regulation

Water supply policies and regulation are usually defined by one or several Ministries, in consultation with the legislative branch. In the United States the United States Environmental Protection Agencyý, whose administrator reports directly to the President, is responsible for water and sanitation policy and standard setting within the executive branch. In other countries responsibility for sector policy is entrusted to a Ministry of Environment (such as in Mexico and Colombia), to a Ministry of Health (such as in Panama, Honduras and Uruguay), a Ministry of Public Works (such as in Ecuador and Haiti), a Ministry of Economy (such as in German states) or a Ministry of Energy (such as in Iran). A few countries, such as Jordan and Bolivia, even have a Ministry of Water. Often several Ministries share responsibilities for water supply. In the European Union, important policy functions have been entrusted to the supranational level. Policy and regulatory functions include the setting of tariff rules and the approval of tariff increases; setting, monitoring and enforcing norms for quality of service and environmental protection; benchmarking the performance of service providers; and reforms in the structure of institutions responsible for service provision. The distinction between policy functions and regulatory functions is not always clear-cut. In some countries they are both entrusted to Ministries, but in others regulatory functions are entrusted to agencies that are separate from Ministries.

Regulatory Agencies

Dozens of countries around the world have established regulatory agencies for infrastructure services, including often water supply and sanitation, in order to better protect consumers and to improve efficiency. Regulatory agencies can be entrusted with a variety of responsibilities, including in particular the approval of tariff increases and the management of sector information systems, including benchmarking systems. Sometimes they also have a mandate to settle complaints by consumers that have not been dealt with satisfactorily by service providers. These specialized entities are expected to be more competent and objective in regulating service providers than departments of government Ministries. Regulatory agencies are supposed to be autonomous from the executive branch of government, but in many countries have often not been able to exercise a great degree of autonomy. In the United States regulatory agencies for utilities have existed for almost a century at the level of states, and in Canada at the level of provinces. In both countries they cover several infrastructure sectors. In many US states they are called Public Utility Commissions. For England and Wales, a regulatory agency for water (OFWAT) was created as part of the privatization of the water industry in 1989. In many developing countries, water regulatory agencies were created during the 1990s in parallel with efforts at increasing private sector participation. (for more details on regulatory agencies in Latin America.

Many countries do not have regulatory agencies for water. In these countries service providers are regulated directly by local government, or the national government. This is, for example, the case in the countries of continental Europe, in China and India.

Service Provision

Water supply service providers, which are often utilities, differ from each other in terms of their geographical coverage relative to administrative boundaries; their sectoral coverage; their ownership structure; and their governance arrangements.

Geographical Coverage

Many water utilities provide services in a single city, town or municipality. However, in many countries municipalities have associated in regional or inter-municipal or multi-jurisdictional utilities to benefit from economies of scale. In the United States these can take the form of special-purpose districts which may have independent taxing authority. An example of a multi-jurisdictional water utility in the United States is WASA, a utility serving Washington, DC and various localities in the state of Maryland. Multi-jurisdictional utilities are also common in Germany, where they are known as "Zweckverbaende", in France and in Italy.

In some federal countries there are water service providers covering most or all cities and towns in an entire state, such as in all states of Brazil and some states in Mexico. In England and Wales water supply and sewerage is supplied almost entirely through ten regional companies. Some smaller countries, especially developed countries, have established service providers that cover the entire country or at least most of its cities and major towns. Such national service providers are especially prevalent in West Africa and Central America, but also exist, for example, in Tunisia, Jordan and Uruguay. In rural areas, where about half the world population lives, water services are often not provided by utilities, but by community-based organizations which usually cover one or sometimes several villages.

Sector Coverage

Some water utilities provide only water supply services, while sewerage is under the responsibility of a different entity. This is for example the case in Tunisia. However, in most cases water utilities also provide sewer and wastewater treatment services. In some cities or countries utilities also distribute electricity. In a few cases such multi-utilities also collect solid waste and provide local telephone services. An example of such an integrated utility can be found in the Colombian city of Medellín. Utilities that provide water, sanitation and electricity can be found in Frankfurt, Germany (Mainova), in Casablanca, Morocco and in Gabon in West Africa. Multi-utilities provide certain benefits such as common billing and the option to cross-subsidize water services with revenues from electricity sales, if permitted by law.

Ownership and Governance Arrangements

Water supply providers can be either public, private, mixed or cooperative. Most urban water supply services around the world are provided by public entities. As Willem-Alexander, Prince of Orange (2002) stated, "The water crisis that is affecting so many people is mainly a crisis of governance — not of water scarcity." The introduction of cost-reflective tariffs together with cross-subsidisation between richer and poorer consumers is an essential governance reform in order to reduce the high levels of Unaccounted or Water (UAW) and to provide the finance needed to extend the network to those poorest households who remain unconnected. Partnership arrangements between the public and private sector can play an important role in order to achieve this objective

Private Sector Participation

An estimated 10 percent of urban water supply is provided by private or mixed public-private companies, usually under concessions, leases or management contracts. Under these arrangements the public entity that is legally responsible for service provision delegates certain or all aspects of service provision to the private service provider for a period typically ranging from 4 to 30 years. The public entity continues to own the assets. These arrangements are common in France and in Spain. Only in few parts of the world water supply systems have been completely sold to the private sector (privatization), such as in England and Wales as well as in Chile. The largest private water companies in the world are Suez and Veolia Environnement from France; Aguas de Barcelona from Spain; and Thames Water from the UK, all of which are engaged internationally.

Public Water Service Provision

90% of urban water supply and sanitation services are currently in the public sector. They are owned by the state or local authorities, or also by collectives or cooperatives. They run without an aim for profit but are based on the ethos of providing a common good considered to be of public interest. In most middle and low-income countries, these publicly-owned and managed water providers can be inefficient as a result of political interference, leading to over-staffing and low labour productivity. Ironically, the main losers from this institutional arrangement are the urban poor in these countries. Because they are not connected to the network, they end up paying far more per litre of water than do more well-off households connected to the network who benefit from the implicit subsidies that they receive from loss-making utilities. We are still so far from achieving universal access to clean water and sanitation shows that public water authorities, in their current state, are not working well enough. Yet some are being very successful and are modelling the best forms of public management. As Ryutaro Hashimoto, former Japanese Prime Minister, notes: "Public water services currently provide more than 90 per cent of water supply in the world. Modest improvement in public water operators will have immense impact on global provision of services."

Governance Arrangements

Governance arrangements for both public and private utilities can take many forms. Governance arrangements define the relationship between the service provider, its owners, its customers and regulatory entities. They determine the financial autonomy of the service provider and thus its ability to maintain its assets, expand services, attract and retain qualified staff, and ultimately to provide high-quality services. Key aspects of governance arrangements are the extent to which the entity in charge of providing services is insulated from arbitrary political intervention; and whether there is an explicit mandate and political will to allow the service provider to recover all or at least most of its costs through tariffs and retain these revenues. If water supply is the responsibility of a department that is integrated in the administration of a city, town or municipality, there is a risk that tariff revenues are diverted for other purposes. In some cases, there is also a risk that staff are appointed mainly on political grounds rather than based on their professional credentials.

Tariffs

Almost all service providers in the world charge tariffs to recover part of their costs. According to estimates by the World Bank the average (mean) global water tariff is US$ 0.53 per cubic meter. In developed countries the average tariff is US$ 1.04, while it is only U$ 0.11 in the poorest developing countries. The lowest tariffs in developing countries are found in South Asia (mean of US$ 0.09/m3), while the highest are found in Latin America (US$ 0.41/m3). Data for 132 cities were assessed. The tariff is estimate for a consumption level of 15 cubic meters per month. Few utilities do recover all their costs. According to the same World Bank study only 30% of utilities globally, and only 50% of utilities in developed countries, generate sufficient revenue to cover operation, maintenance and partial capital costs.

According to another study undertaken in 2006 by NUS Consulting, the average water and sewerage tariff in 14 mainly OECD countries excluding VAT varied between US$ 0.66 per cubic meter in the United States and the equivalent of US$ 2.25 per cubic meter in Denmark. However, water consumption is much higher in the US than in Europe. Therefore, residential water bills may be very similar, even if the tariff per unit of consumption tends to be higher in Europe than in the US. A typical family on the US East Coast paid between US$30 and US$70 per month for water and sewer services in 2005.

In developing countries tariffs are usually much further from covering costs. Residential water bills for a typical consumption of 15 cubic meters per month vary between less than US$ 1 and US$ 12 per month. Water and sanitation tariffs, which are almost always billed together, can take many different forms. Where meters are installed, tariffs are typically volumetric (per usage), sometimes combined with a small monthly fixed charge. In the absence of meters, flat or fixed rates — which are independent of actual consumption — are being charged. In developed countries, tariffs are usually the same for different categories of users and for different levels of consumption.

In developing countries, the situation is often characterized by cross-subsidies with the intent to make water more affordable for residential low-volume users that are assumed to be poor. For example, industrial and commercial users are often charged higher tariffs than public or residential users. Also, metered users are often charged higher tariffs for higher levels of consumption (increasing-block tariffs). However, cross-subsidies between residential users do not always reach their objective. Given the overall low level of water tariffs in developing countries even at higher levels of consumption, most consumption subsidies benefit the wealthier segments of society. Also, high industrial and commercial tariffs can provide an incentive for these users to supply water from other sources than the utility (own wells, water tankers) and thus actually erode the utility's revenue base.

Water Meter

A water meter is a device used to measure the volume of water usage. This article provides an overview of technical aspects of water meters. The worldwide prevalence of metering as well as its economic benefits and costs are covered in the separate article on water metering.

In many developed countries water meters are used at each residential and commercial building in a public water supply system. Water meters can also be used at the water source, well, or throughout a water system to determine flow through that portion of the system. Water meters typically measure and display total usage in cubic feet (ft.^{3}), cubic meters (m^3) or US gallons on a mechanical or electronic register. Some electronic meter registers can display rate-of-flow as well as total usage. There are several types of water meter in common use. Selection is based on different flow measurement methods, the type of end user, the required flow rates, and accuracy requirements.

In North America, standards for manufacturing of water meters are made by the American Water Works Association.

Types of Meters

There are two major methods of flow measurement in use, displacement and velocity, with sub-technologies within each of them. Common displacement designs include oscillating piston and nutating disk meters. Velocity-based designs include single- and multi-jet meters and turbine meters. There are also non-mechanical designs, for example electromagnetic and ultrasonic meters.

In addition to the more common types of meter, there are meters designed for special uses. Most meters in a typical water distribution system are designed for cold potable water only. There are specialty water meters manufactured for specific other uses. Hot water meters are designed with special materials that can withstand higher temperatures. Meters for reclaimed water have special lavender register covers to signify that the water is non-potable and should not be used for drinking.

Water meters are generally owned, read, and maintained by a public water provider such as a city, rural water association, or private water company. In some cases an owner of a mobile home park, apartment complex or commercial building may be billed by a utility on one meter, and want to share the cost of the bill among the tenants. In this case, private meters may be purchased to separately track usage of each unit in what is called submetering.

Displacement Water Meters

This type of water meter is most often used in residential and small commercial applications. Displacement meters are commonly referred to as Positive Displacement, or "PD" meters. Two common methods of positive displacement measuring are Oscillating Piston meters and Nutating Disk meters. Either method relies on the water to physically displace the moving measuring element in direct relation to the amount of water that passes through the meter. The piston or disk moves a magnet that drives the register.

Figure: *Water meter in Belo Horizonte.*

PD meters are generally very accurate at low to moderate flow rates typical of residential and small commercial users, and are common in sizes from 5/8" to 2". Because displacement meters rely on all water flowing through the meter to "push" the measuring element, they generally are not practical in large commercial applications requiring high flow rates or low pressure loss. PD meters normally have a built-in strainer to protect the measuring element from rocks or other debris that could stop or break the measuring element. PD meters normally have bronze, brass or plastic bodies with internal measuring chambers made from molded plastics and stainless steel.

Velocity Water Meters

A velocity-type meter measures the velocity of flow through a meter of a known internal capacity. The speed of the flow can then be converted into volume of flow for usage. There are several types of meters that measure water flow velocity to determine totality usage. They include jet meters (single-jet and multi-jet), turbine meters, propeller meters, and mag meters. Most velocity-based meters have an adjustment vane for calibration of the meter to required accuracy.

Multi-jet Meters

Multi-jet meters are very accurate in small sizes and are commonly used in 5/8" to 2" sizes for residential and smaller commercial uses. Multi-jet meters use multiple ports surrounding an internal chamber, to create multiple jets of water against an impeller. The impeller rotation speed is in relation to the velocity of water flow. Multi-jets are very accurate at low flow rates, but are not used in larger sizes, since they don't have the straight-through flow path needed for the high flow rates used in large pipe diameters. Multi-jet meters generally have an internal strainer element that can protect the jet ports from getting clogged. Multi-jet meters normally have bronze alloy bodies or outer casings, with internal measuring parts made from modern thermoplastics and stainless steel.

Turbine Meters

Turbine meters are less accurate than displacement and jet meters at low flow rates, but the measuring element does not occupy or severely restrict the entire path of flow. The flow direction is generally straight through the meter, allowing for higher flow rates and less pressure loss than displacement-type meters. They are the meter of choice for large commercial users, fire protection, and as master meters for the water distribution system. Strainers are generally required to be installed in front of the meter to protect the measuring element from gravel or other debris that could enter the water distribution system. Turbine meters are generally available for 1-1/2" to 12" or higher pipe sizes. Turbine meter bodies are commonly made of bronze, cast Iron, or ductile iron. Internal turbine elements can be plastic or non-corrosive metal alloys. they are accurate in normal working conditions to 0.2l/s however are affect greatly with dog mix interference.

- Fire meters are a specialized type of turbine meter with approvals for the high flow rates required in fire protection. They are often approved by Underwriters Laboratories (UL) or Factory Mutual (FM) for use in fire protection.
- Fire hydrant meters are a specialized type of portable turbine meter that are attached to a fire hydrant to measure water out of the hydrant. The meters are normally made of aluminum to be light weight, and are usually 3" capacity. Utilities often

require them for measurement of water used in construction, pool filling, or where a permanent meter is not yet installed.

Compound Meters

A compound meter is used where high flow rates are necessary, but at times there are smaller rates of flow that still need to be accurately measured. Compound meters have two measuring elements and a check valve to regulate flow between them. At high flow rates, water is normally diverted primarily or completely to the turbine part of the meter. When flow rates drop to where the turbine meter cannot measure accurately, a check valve closes to divert water to a smaller meter that can measure the lower flow rates accurately. The low flow meter is typically a multi-jet or PD meter. By adding the registration of the high and low meter registers, the utility has the total consumption through the meter.

Electromagnetic Meters

Figure: *Electromagnetic flow meter*

Magnetic flow meters, commonly referred to as "mag meters", are technically a velocity-type water meter, except that they use electromagnetic properties to determine the water flow velocity, rather than mechanical means which jet and turbine meters use. Mag meters use the physics principle of Faraday's law of induction for measurement, and require AC or DC electricity from line or battery to operate the electromagnets. Since mag meters have no mechanical measuring element, they normally have the advantage of being able to measure flow in either direction, and use electronics for measuring and totalizing the flow. Mag meters can also be useful for measuring untreated water, raw (untreated/unfiltered) water, and wastewater, since there is no mechanical measuring element to get clogged or damaged by debris flowing through the meter. Strainers are not required with mag meters, since there is no measuring element in the stream of flow that could be damaged. Stray electrical energy flowing through the flow tube can cause inaccurate readings, therefore most mag meters are installed with either grounding rings or grounding electrodes to divert stray electricity away from the electrodes inside the flow tube which are used to measure the flow.

Ultrasonic Meters

Ultrasonic water meters use an ultrasonic transducer to send ultrasonic sound waves through the fluid to determine the velocity and translate the velocity into measurement of the water volume.

Registers

Figure: *A typical water meter register showing a meter reading of 8.3 gallons. Notice the black "1" on the odometer has not yet fully turned over, so the red hand is read in its place.*

There are several types of registers on water meters. A standard register normally has a dial similar to a clock, with gradations around the perimeter to indicate water usage measured by the meter, as well as a set of odometer wheels similar to that in a car. Modern registers are normally driven by a magnetic coupling between a magnet in the measuring chamber attached to the measuring element, and another attached to the bottom of the register. Gears in the register convert the motion of the measuring element to the proper usage increment for display on the sweep hand and the odometer. Many registers also have a leak detector. This is a small visible disk or hand that is geared closer to the rotation speed of the drive magnet, so that very small flows that would be visually undetectable on the regular sweep hand can be seen. With Automatic Meter Reading, manufacturers have developed pulse or encoder registers to produce electronic output for radio transmitters, reading storage devices, and data logging devices. Pulse meters send a digital or analog electronic pulse to a recording device. Encoder registers have an electronic means for an external device to interrogate the register for either the position of the odometer wheels or a stored electronic reading. Frequent transmissions of consumption can be used to give smart meter functionality.

There are also some specialized types of registers such as LCD display instead of mechanical odometers, and registers to output data or pulses to a variety of recording and controller devices. For industrial applications, output is often 4-20 mA analog, for the recording or control of different flow rates in addition to totalization.

Water Meter Reading

Different size meters may indicate a reading in different resolutions. One rotation of the sweep hand may be 10 gallons or 1000 gallons (1 to 100 ft.^{3}, 0.1 to 10 m^3). If one rotation of the hand represents 10 gallons, the meter has a 10 gallon sweep. sometimes the last number(s) of the odometer are non-rotating or printed on the dial face. The fixed zero number(s) are represented by the position of the rotating sweep hand. If one rotation of the hand is 10 gallons, the sweep hand is on 7, and the odometer shows 123456 plus a fixed zero, the actual total usage would be 1,234,567 gallons. A moving odometer wheel that has not completely turned over is counted as the lower number. In the United States most utilities bill only to the nearest 100 or 1000 gallons (10 to 100 ft.^{3}, 1 to 10 m^3), and often only read the leftmost 4 or 5 numbers on the odometer. They would only read and bill using 1234 and rounding to 1,234,000 gallons based on a 1000 gallon billing resolution and the example above. The most common rounding for a particular size meter will often have different colored number wheels, the ones ignored being black, and the ones used for billing being white.

Metering

Metering of water supply is usually motivated by one or several of four objectives: First, it provides an incentive to conserve water which protects water resources (environmental objective). Second, it can postpone costly system expansion and saves energy and chemical costs (economic objective). Third, it allows a utility to better locate distribution losses (technical objective). Fourth, it allows to charge for water based on use, which is perceived by many as the fairest way to allocate the costs of water supply to users. Metering is considered good practice in water supply and is widespread in developed countries, except for the United Kingdom. In developing countries it is estimated that half of all urban water supply systems are metered and the tendency is increasing.

Water meters are read by one of several methods:

- the water customer writes down the meter reading and mails in a postcard with this info to the water department;
- the water customer writes down the meter reading and uses a phone dial-in system to transfer this info to the water department;
- the water customer logs in to the website of the water supply company, enters the address, meter ID and meter readings [6]
- a meter reader comes to the premise and enters the meter reading into a handheld computer;
- the meter reading is echoed on a display unit mounted to the outside of the premise, where a meter reader records them;
- a small radio is hooked up to the meter to automatically transmit readings to corresponding receivers in handheld computers, utility vehicles or distributed collectors
- a small computer is hooked up to the meter that can either dial out or receive automated phone calls that give the reading to a central computer system.

Most cities are increasingly installing Automatic Meter Reading (AMR) systems to prevent fraud, to lower ever-increasing labor and liability costs and to improve customer service and satisfaction.

Costs and Financing

The cost of supplying water consists to a very large extent of fixed costs (capital costs and personnel costs) and only to a small extent of variable costs that depend on the amount of water consumed (mainly energy and chemicals). The full cost of supplying water in urban areas in developed countries is about US$1–2 per cubic meter depending on local costs and local water consumption levels. The cost of sanitation (sewerage and wastewater treatment) is another US$1–2 per cubic meter. These costs are somewhat lower in developing countries. Throughout the world, only part of these costs is usually billed to consumers, the remainder being financed through direct or indirect subsidies from local, regional or national governments.

Besides subsidies water supply investments are financed through internally generated revenues as well as through debt. Debt financing can take the form of credits from commercial Banks, credits from international financial institutions such as the World Bank and regional development banks (in the case of developing countries), and bonds (in the case of some developed countries and some upper middle-income countries).

History

Throughout history people have devised systems to make getting and using water more convenient. Early Rome had indoor plumbing, meaning a system of aqueducts and pipes that terminated in homes and at public wells and fountains for people to use. London water supply infrastructure developed over many centuries from early mediaeval conduits, through major 19th century treatment works built in response to cholera threats, to modern large scale reservoirs.

Water towers appeared around the late 19th century, as building height rose, and steam, electric and diesel-powered water pumps became available. As skyscrapers appeared, they needed rooftop water towers. The technique of purification of drinking water by use of compressed liquefied chlorine gas was developed in 1910 by U.S. Army Major (later Brig. Gen.) Carl Rogers Darnall (1867–1941), Professor of Chemistry at the Army Medical School. Shortly thereafter, Major (later Col.) William J. L. Lyster (1869–1947) of the Army Medical Department used a solution of calcium hypochlorite in a linen bag to treat water. For many decades, Lyster's method remained the standard for U.S. ground forces in the field and in camps, implemented in the form of the familiar Lyster Bag (also spelled Lister Bag). Darnall's work became the basis for present day systems of municipal water '*purification*'.

Desalination appeared during the late 20th century, and is still limited to a few areas.

During the beginning of the 21st Century, especially in areas of urban and suburban population centres, traditional centralized infrastructure have not been able to supply sufficient quantities of water to keep up with growing demand. Among several options that have been managed are the extensive use of desalination technology, this is especially prevalent in coastal areas and in "dry" countries like Australia. Decentralization of water infrastructure has grown extensively as a viable solution including Rainwater harvesting and Stormwater harvesting where policies are eventually tending towards a more rational use and sourcing of water incorporation concepts such as "Fit for Purpose".

Standardization

International standards for water supply system are covered by International Classification of Standards (ICS) 91.140.60.

Outbreaks of Diseases Due to Contaminated Water Supply

- In 1854, a cholera outbreak in London's Soho district was identified by Dr. John Snow as originating from contaminated water from the Broad street pump. This can be regarded as a founding event of the science of epidemiology.
- In 1980, a hepatitis A surge due to the consumption of water from a feces-contaminated well, in Pennsylvania
- In 1987, a cryptosporidiosis outbreak is caused by the public water supply of which the filtration was contaminated, in western Georgia
- Fluoride intoxication in a long-term hemodialysis unit of university hospital due to the failure of a water deionization system
- In 1988, many people were poisoned in Camelford, when a worker put 20 tonnes of aluminium sulphate in the wrong tank.
- In 1993, a fluoride poisoning outbreak resulting from overfeeding of fluoride, in Mississippi
- In 1993, Milwaukee Cryptosporidium outbreak
- An outbreak of typhoid fever in northern Israel, which was associated with the contaminated municipal water supply
- In 1997, 369 cases of cryptosporidiosis occurred, caused by a contaminated fountain in the Minnesota zoo. Most of the sufferers were children
- In 1998, a non-chlorinated municipal water supply was blamed for a campylobacteriosis outbreak in northern Finland

- In 2000, a gastroenteritis outbreak that was brought by a non-chlorinated community water supply, in southern Finland
- In 2000, an E. coli outbreak occurred in Walkerton Ontario Canada. Seven people died from drinking contaminated water. Hundreds suffered from the symptoms of the disease, not knowing if they too would die.
- In 2004, contamination of the community water supply, serving the Bergen city centre of Norway, was later reported after the outbreak of waterborne giardiasis
- In 2007, contaminated drinking water was pinpointed which had led to the outbreak of gastroenteritis with multiple aetiologies in Denmark

Agriculture: The most important use of water in agriculture is for irrigation, which is a key component to produce enough food. Irrigation takes up to 90% of water withdrawn in some developing countries and significant proportions in more economically developed countries (United States, 30% of freshwater usage is for irrigation). It takes around 3,000 litres of water, converted from liquid to vapour, to produce enough food to satisfy one person's daily dietary need. This is a considerable amount, when compared to that required for drinking, which is between two and five litres. To produce food for the 6.5 billion or so people who inhabit the planet today requires the water that would fill a canal ten metres deep, 100 metres wide and 7.1 million kilometres long – that's enough to circle the globe 180 times.

Fifty years ago, the common perception was that water was an infinite resource. At this time, there were fewer than half the current number of people on the planet. People were not as wealthy as today, consumed fewer calories and ate less meat, so less water was needed to produce their food. They required a third of the volume of water we presently take from rivers. Today, the competition for the fixed amount of water resources is much more intense, giving rise to the concept of peak water. This is because there are now nearly seven billion people on the planet, their consumption of water-thirsty meat and vegetables is rising, and there is increasing competition for water from industry, urbanisation and biofuel crops. In future, even more water will be needed to produce food because the Earth's population is forecast to rise to 9 billion by 2050. An additional 2.5 or 3 billion people, choosing to eat fewer cereals and more meat and vegetables could add an additional five million kilometres to the virtual canal mentioned above.

An assessment of water management in agriculture was conducted in 2007 by the International Water Management Institute in Sri Lanka to see if the world had sufficient water to provide food for its growing population. It assessed the current availability of water for agriculture on a global scale and mapped out locations suffering from water scarcity. It found that a fifth of the world's people, more than 1.2 billion, live in areas of physical water scarcity, where there is not enough water to meet all demands. A further 1.6 billion people live in areas experiencing economic water scarcity, where the lack of investment in water or insufficient human capacity make it impossible for authorities to satisfy the demand for water. The report found that it would be possible to produce the food required in future, but that continuation of today's food production and environmental trends would lead to crises in many parts of the world. To avoid a global water crisis, farmers will have to strive to increase productivity to meet growing demands for food, while industry and cities find ways to use water more efficiently.

As a Scientific Standard

On 7 April 1795, the gram was defined in France to be equal to "the absolute weight of a volume of pure water equal to a cube of one hundredth of a meter, and to the temperature of the melting ice." For practical purposes though, a metallic reference standard was required, one thousand times more massive, the kilogram. Work was therefore commissioned to determine precisely the mass of one liter of water. In spite of the fact that the decreed definition of the gram specified water at 0 °C — a highly reproducible *temperature* — the scientists chose to redefine the standard and to perform their measurements at the temperature of highest water *density*, which was measured at the time as 4 °C (39 °F).

The Kelvin temperature scale of the SI system is based on the triple point of water, defined as exactly 273.16 K or 0.01 °C. The scale is an absolute temperature scale with the same increment as the Celsius temperature scale, which was originally defined according the boiling point (set to 100 °C) and melting point (set to 0 °C) of water.

Natural water consists mainly of the isotopes hydrogen-1 and oxygen-16, but there is also small quantity of heavier isotopes such as hydrogen-2 (deuterium). The amount of deuterium oxides or heavy water is very small, but it still affects the properties of water. Water from rivers and lakes tends to contain less deuterium than seawater.

Therefore, standard water is defined in the Vienna Standard Mean Ocean Water specification.

For Drinking

The human body contains from 55% to 78% water, depending on body size. To function properly, the body requires between one and seven liters of water per day to avoid dehydration; the precise amount depends on the level of activity, temperature, humidity, and other factors. Most of this is ingested through foods or beverages other than drinking straight water. It is not clear how much water intake is needed by healthy people, though most advocates agree that approximately 2 liters (6 to 7 glasses) of water daily is the minimum to maintain proper hydration.

Medical literature favors a lower consumption, typically 1 liter of water for an average male, excluding extra requirements due to fluid loss from exercise or warm weather. For those who have healthy kidneys, it is rather difficult to drink too much water, but (especially in warm humid weather and while exercising) it is dangerous to drink too little. People can drink far more water than necessary while exercising, however, putting them at risk of water intoxication (hyperhydration), which can be fatal. The popular claim that "a person should consume eight glasses of water per day" seems to have no real basis in science. Similar misconceptions concerning the effect of water on weight loss and constipation have also been dispelled.

An original recommendation for water intake in 1945 by the Food and Nutrition Board of the United States National Research Council read: "An ordinary standard for diverse persons is 1 milliliter for each calorie of food. Most of this quantity is contained in prepared foods." The latest dietary reference intake report by the United States National Research Council in general recommended (including food sources): 3.7 liters for men and 2.7 liters of water total for women. Specifically, pregnant and breastfeeding women need additional fluids to stay hydrated. The Institute of Medicine (U.S.) recommends that, on average, men consume 3.0 liters and women 2.2 liters; pregnant women should increase intake to 2.4 liters (10 cups) and breastfeeding women should get 3 liters (12 cups), since an especially large amount of fluid is lost during nursing. Also noted is that normally, about 20% of water intake comes from food, while the rest comes from drinking water and beverages (caffeinated included). Water is excreted from the body in multiple forms; through urine and feces, through sweating, and by exhalation of water vapor in the breath. With physical exertion and heat exposure, water loss will increase and daily fluid needs may increase as well.

Humans require water with few impurities. Common impurities include metal salts and oxides, including copper, iron, calcium and lead, and/or harmful bacteria, such as *Vibrio*. Some solutes are acceptable and even desirable for taste enhancement and to provide needed electrolytes.

The single largest (by volume) freshwater resource suitable for drinking is Lake Baikal in Siberia.

Washing

The propensity of water to form solutions and emulsions is useful in various washing processes. Many industrial processes rely on reactions using chemicals dissolved in water, suspension of solids in water slurries or using water to dissolve and extract substances. Washing is also an important component of several aspects of personal body hygiene.

Transportation

The use of water for transportation of materials through rivers and canals as well as the international shipping lanes is an important part of the world economy.

Chemical Uses

Water is widely used in chemical reactions as a solvent or reactant and less commonly as a solute or catalyst. In inorganic reactions, water is a common solvent, dissolving many ionic compounds. In organic reactions, it is not usually used as a reaction solvent, because it does not dissolve the reactants well and is amphoteric (acidic *and* basic) and nucleophilic. Nevertheless, these properties are sometimes desirable. Also, acceleration of Diels-Alder reactions by water has been observed. Supercritical water has recently been a topic of research. Oxygen-saturated supercritical water combusts organic pollutants efficiently.

Heat Exchange

Water and steam are used as heat transfer fluids in diverse heat exchange systems, due to its availability and high heat capacity, both as a coolant and for heating. Cool water may even be naturally available from a lake or the sea. Condensing steam is a particularly efficient heating fluid because of the large heat of vaporization. A disadvantage is that water and steam are somewhat corrosive. In almost all electric power stations, water is the coolant, which vaporizes and drives steam turbines to drive generators. In the U.S., cooling power plants is the largest use of water.

In the nuclear power industry, water can also be used as a neutron moderator. In most nuclear reactors, water is both a coolant and a moderator. This provides something of a passive safety measure, as removing the water from the reactor also slows the nuclear reaction down – however other methods are favored for stopping a reaction and it is preferred to keep the nuclear core covered with water so as to ensure adequate cooling.

Fire Extinction

Water has a high heat of vaporization and is relatively inert, which makes it a good fire extinguishing fluid. The evaporation of water carries heat away from the fire. It is dangerous to use water on fires involving oils and organic solvents, because many organic materials float on water and the water tends to spread the burning liquid. Use of water in fire fighting should also take into account the hazards of a steam explosion, which may occur when water is used on very hot fires in confined spaces, and of a hydrogen explosion, when substances which react with water, such as certain metals or hot carbon such as coal, charcoal, coke graphite, decompose the water, producing water gas. The power of such explosions was seen in the Chernobyl disaster, although the water involved did not come from fire-fighting at that time but the reactor's own water cooling system. A steam explosion occurred when the extreme overheating of the core caused water to flash into steam. A hydrogen explosion may have occurred as a result of reaction between steam and hot zirconium.

Recreation

Humans use water for many recreational purposes, as well as for exercising and for sports. Some of these include swimming, waterskiing, boating, surfing and diving. In addition, some sports, like ice hockey and ice skating, are played on ice. Lakesides, beaches and waterparks are popular places for people to go to relax and enjoy recreation. Many find the sound and appearance of flowing water to be calming, and fountains and other water features are popular decorations. Some keep fish and other life in aquariums or ponds for show, fun, and companionship. Humans also use water for snow sports i.e. skiing, sledding, snowmobiling or snowboarding, which requires the water to be frozen.

Water Industry

The water industry provides drinking water and wastewater services (including sewage treatment) to households and industry. Water supply facilities include water wells cisterns for rainwater harvesting, water supply network, water purification facilities, water tanks, water towers, water pipes including old aqueducts. Atmospheric water generators are in development. Drinking water is often collected at springs, extracted from artificial borings (wells) in the ground, or pumped from lakes and rivers. Building more wells in adequate places is thus a possible way to produce more water, assuming the aquifers can supply an adequate flow. Other water sources include rainwater collection. Water may require purification for human consumption. This may involve removal of undissolved substances, dissolved substances and harmful microbes. Popular methods are filtering with sand which only removes undissolved material, while chlorination and boiling kill harmful microbes. Distillation does all three functions. More advanced techniques exist, such as reverse osmosis. Desalination of abundant seawater is a more expensive solution used in coastal arid climates. The distribution of drinking water is done through municipal water systems, tanker delivery or as bottled water. Governments in many countries have programs to distribute water to the needy at no charge.

Reducing usage by using drinking (potable) water only for human consumption is another option. In some cities such as Hong Kong, sea water is extensively used for flushing toilets citywide in order to conserve fresh water resources. Polluting water may be the biggest single misuse of water; to the extent that a pollutant limits other uses of the water, it becomes a waste of the resource, regardless of benefits to the polluter. Like other types of pollution, this does not enter standard accounting of market costs, being conceived as externalities for which the market cannot account. Thus other people pay the price of water pollution, while the private firms' profits are not redistributed to the local population victim of this pollution. Pharmaceuticals consumed by humans often end up in the waterways and can have detrimental effects on aquatic life if they bioaccumulate and if they are not biodegradable. Wastewater facilities are storm sewers and wastewater treatment plants. Another way to remove pollution from surface runoff water is bioswale.

Industrial Applications

Water is used in power generation. Hydroelectricity is electricity obtained from hydropower. Hydroelectric power comes from water driving a water turbine connected to a generator. Hydroelectricity is a low-cost, non-polluting, renewable energy source. The energy is supplied by the motion of water. Typically a dam is constructed on a river, creating an artificial lake behind it. Water flowing out of the lake is forced through turbines that turn generators. Pressurized water is used in water blasting and water jet cutters. Also, very high pressure water guns are used for precise cutting. It works very well, is relatively safe, and is not harmful to the environment. It is also used in the cooling of machinery to prevent overheating, or prevent saw blades from overheating. Water is also used in many industrial processes and machines, such as the steam turbine and heat exchanger, in addition to its use as a chemical solvent. Discharge of untreated water from industrial uses is pollution. Pollution includes discharged solutes (chemical pollution) and discharged coolant water (thermal pollution). Industry requires pure water for many applications and utilizes a variety of purification techniques both in water supply and discharge.

Food Processing

Water plays many critical roles within the field of food science. It is important for a food scientist to understand the roles that water plays within food processing to ensure the success of their products.

Solutes such as salts and sugars found in water affect the physical properties of water. The boiling and freezing points of water are affected by solutes, as well as air pressure, which is in turn affected by altitude. Water boils at lower temperatures with the lower air pressure which occurs at higher elevations. One mole of sucrose (sugar) per kilogram of water raises the boiling point of water by 0.51 °C, and one mole of salt per kg raises the boiling point by 1.02 °C; similarly, increasing the number of dissolved particles lowers water's freezing point. Solutes in water also affect water activity which affects many chemical reactions and the growth of microbes in food. Water activity can be described as a ratio of the vapor pressure of water in a solution to the vapor pressure of pure water. Solutes in water lower water activity. This is important to know because most bacterial growth ceases at low levels of water activity. Not only does microbial growth affect the safety of food but also the preservation and shelf life of food.

Water hardness is also a critical factor in food processing. It can dramatically affect the quality of a product as well as playing a role in sanitation. Water hardness is classified based on the amounts of removable calcium carbonate salt it contains per gallon. Water hardness is measured in grains; 0.064 g calcium carbonate is equivalent to one grain of hardness. Water is classified as soft if it contains 1 to 4 grains, medium if it contains 5 to 10 grains and hard if it contains 11 to 20 grains. The hardness of water may be altered or treated by using a chemical ion exchange system. The hardness of water also affects its pH balance which plays a critical role in food processing. For example, hard water prevents successful production of clear beverages. Water hardness also affects sanitation; with increasing hardness, there is a loss of effectiveness for its use as a sanitizer.

Boiling, steaming, and simmering are popular cooking methods that often require immersing food in water or its gaseous state, steam. Water is also used for dishwashing.

13

Soil Contamination

Soil contamination or soil pollution is caused by the presence of xenobiotic (human-made) chemicals or other alteration in the natural soil environment.

This type of contamination typically arises from the failure caused by corrosion of underground storage tanks (including piping used to transmit the contents), application of pesticides, percolation of contaminated surface water to subsurface strata, oil and fuel dumping, disposal of coal ash, leaching of wastes from landfills or direct discharge of industrial wastes to the soil. The most common chemicals involved are petroleum hydrocarbons, lead, polynuclear aromatic hydrocarbons (such as naphthalene and benzo(a)pyrene), solvents, pesticides, and other heavy metals. This occurrence of this phenomenon is correlated with the degree of industrialization and intensities of chemical usage.

The concern over soil contamination stems primarily from health risks, from direct contact with the contaminated soil, vapors from the contaminants, and from secondary contamination of water supplies within and underlying the soil. Mapping of contaminated soil sites and the resulting cleanup are time consuming and expensive tasks, requiring extensive amounts of geology, hydrology, chemistry, computer modeling skills, and GIS in Environmental Contamination, as well as an appreciation of the history of industrial chemistry.

It is in North America and Western Europe that the extent of contaminated land is most well known, with many of countries in these areas having a legal framework to identify and deal with this environmental problem; this however may well be just the tip of the iceberg with developing countries very likely to be the next generation of new soil contamination cases.

The immense and sustained growth of the People's Republic of China since the 1970s has exacted a price from the land in increased soil pollution. The State Environmental Protection Administration believes it to be a threat to the environment, to food safety and to sustainable agriculture. According to a scientific sampling, 150 million mi (100,000 square kilometers) of China's cultivated land have been polluted, with contaminated water being used to irrigate a further 32.5 million mi (21,670 square kilometers) and another 2 million mi (1,300 square kilometers) covered or destroyed by solid waste. In total, the area accounts for one-tenth of China's cultivatable land, and is mostly in economically developed areas. An estimated 12 million tonnes of grain are contaminated by heavy metals every year, causing direct losses of 20 billion yuan (US$2.57 billion).

Causes

This type of contamination or pollution typically arises from failure due to corrosion of underground storage tanks or of the piping associated with them, historical disposal of coal ash, application of pesticides, percolation of contaminated surface water to subsurface strata, oil and fuel dumping, leaching of wastes from landfills or direct discharge of industrial wastes to the soil. The most common chemicals involved are petroleum hydrocarbons, solvents, lead, pesticides, and other heavy metals. This occurrence of this phenomenon is correlated with the degree of industrialization and intensities of chemical usage.

Hitorical deposition of coal ash used for residential, commercial, and industrial heating, as well as for industrial processes such as ore smelting, is a common source of contamination in areas that were industrialized before about 1960. Coal natually concentrates lead and zinc during its formation, as well as other heavy metals to a lesser degree. When the coal is burned, most of these metals become concentrated in the ash (the principal exception being mercury). Coal ash and slag may contain sufficient lead to qualify as a "characteristic hazardous waste", defined in the USA as containing more than 5 mg/L of extractable lead using the TCLP procedure. In addition to lead, coal ash typically contains variable but significant concentrations of polynuclear aromatic hydrocarbons (PAHs; e.g., benzo(a)anthracene, benzo(b)fluoranthene, benzo(k)fluoranthene, benzo(a)pyrene, indeno(cd)pyrene, phenanthrene, anthracene, and others). These PAHs are known human carcinogens and the acceptable concentrations of them in soil are typically around 1 mg/kg. Coal ash and slag can be recognized by the presence of off-white grains in soil, gray heterogeneous soil, or (coal slag) bubbly, vesicular pebble-sized grains.

Treated sewage sludge, known in the industry as biosolids, has become controversial as a fertilizer to the land. As it is the byproduct of sewage treatment, it generally contains contaminants such as organisms, pesticides, and heavy metals than other soil.

Health Effects

Contaminated or polluted soil directly affects human health through direct contact with soil or via inhalation of soil contaminants which have vaporized; potentially greater threats are posed by the infiltration of soil contamination into groundwater aquifers used for human consumption, sometimes in areas apparently far removed from any apparent source of above ground contamination.

Health consequences from exposure to soil contamination vary greatly depending on pollutant type, pathway of attack and vulnerability of the exposed population. Chronic exposure to chromium, lead and other metals, petroleum, solvents, and many pesticide and herbicide formulations can be carcinogenic, can cause congenital disorders, or can cause other chronic health conditions. Industrial or man-made concentrations of naturally-occurring substances, such as nitrate and ammonia associated with livestock manure from agricultural operations, have also been identified as health hazards in soil and groundwater.

Chronic exposure to benzene at sufficient concentrations is known to be associated with higher incidence of leukemia. Mercury and cyclodienes are known to induce higher incidences of kidney damage, some irreversible. PCBs and cyclodienes are linked to liver toxicity. Organophosphates and carbamates can induce a chain of responses leading to neuromuscular blockage. Many chlorinated solvents induce liver changes, kidney changes and depression of the central nervous system. There is an entire spectrum of further health effects such as headache, nausea, fatigue, eye irritation and skin rash for the above cited and other chemicals. At sufficient dosages a large number of soil contaminants can cause death by exposure via direct contact, inhalation or ingestion of contaminants in groundwater contaminated through soil.

Ecosystem Effects

Not unexpectedly, soil contaminants can have significant deleterious consequences for ecosystems. There are radical soil chemistry changes which can arise from the presence of many hazardous chemicals even at low concentration of the contaminant species. These changes can manifest in the alteration of metabolism of endemic microorganisms and arthropods resident in a given soil environment. The result can be virtual eradication of some of the primary food chain, which in turn could have major consequences for predator or consumer species. Even if the chemical effect on lower life forms is small, the lower pyramid levels of the food chain may ingest alien chemicals, which normally become more concentrated for each consuming rung of the food chain. Many of these effects are now well known, such as the concentration of persistent DDT materials for avian consumers, leading to weakening of egg shells, increased chick mortality and potential extinction of species.

Effects occur to agricultural lands which have certain types of soil contamination. Contaminants typically alter plant metabolism, most commonly to reduce crop yields. This has a secondary effect upon soil conservation, since the languishing crops cannot shield the Earth's soil mantle from erosion phenomena. Some of these chemical contaminants have long half-lives and in other cases derivative chemicals are formed from decay of primary soil contaminants.

Cleanup Options

Clean up or environmental remediation is analyzed by environmental scientists who utilize field measurement of soil chemicals and also apply computer models (GIS in Environmental Contamination) for analyzing transport and fate of soil chemicals. There are several principal strategies for remediation:

- Excavate soil and take it to a disposal site away from ready pathways for human or sensitive ecosystem contact. This technique also applies to dredging of bay muds containing toxins.
- Aeration of soils at the contaminated site (with attendant risk of creating air pollution)
- Thermal remediation by introduction of heat to raise subsurface temperatures sufficiently high to volatize chemical contaminants out of the soil for vapour extraction. Technologies include ISTD, electrical resistance heating (ERH), and ET-DSP.

- Bioremediation, involving microbial digestion of certain organic chemicals. Techniques used in bioremediation include landfarming, biostimulation and bioaugmentating soil biota with commercially available microflora.
- Extraction of groundwater or soil vapor with an active electromechanical system, with subsequent stripping of the contaminants from the extract.
- Containment of the soil contaminants (such as by capping or paving over in place).
- Phytoremediation, or using plants (such as willow) to extract heavy metals

Light Pollution

Light pollution, also known as photopollution or luminous pollution, is excessive or obtrusive artificial light. The International Dark-Sky Association (IDA) defines *light pollution* as:

Any adverse effect of artificial light including sky glow, glare, light trespass, light clutter, decreased visibility at night, and energy waste. This approach confuses the cause and the result, however. Pollution is the adding-of/added light itself, in analogy to added sound, carbon dioxide, etc. Adverse consequences are multiple; some of them may be not known yet. Scientific definitions thus include the following:

- Alteration of natural light levels in the outdoor environment owing to artificial light sources.
- Light pollution is the alteration of light levels in the outdoor environment (from those present naturally) due to man-made sources of light. Indoor light pollution is such alteration of light levels in the indoor environment due to sources of light, which compromises human health.
- Light pollution is the introduction by humans, directly or indirectly, of artificial light into the environment.

The first two of the above three scientific definitions describe the state of the environment. The third (and newest) one describes the process of polluting by light.

Light pollution obscures the stars in the night sky for city dwellers, interferes with astronomical observatories, and, like any other form of pollution, disrupts ecosystems and has adverse health effects. Light pollution can be divided into two main types:

(1) annoying light that intrudes on an otherwise natural or low-light setting and

(2) excessive light (generally indoors) that leads to discomfort and adverse health effects. Since the early 1980s, a global dark-sky movement has emerged, with concerned people campaigning to reduce the amount of light pollution.

Light pollution is a side effect of industrial civilization. Its sources include building exterior and interior lighting, advertising, commercial properties, offices, factories, streetlights, and illuminated sporting venues. It is most severe in highly industrialized, densely populated areas of North America, Europe, and Japan and in major cities in the Middle East and North Africa like Tehran and Cairo, but even relatively small amounts of light can be noticed and create problems. Like other forms of pollution (such as air, water, and noise pollution) light pollution causes damage to the environment.

Impact on Energy Usage

Energy conservation advocates contend that light pollution must be addressed by changing the habits of society, so that lighting is used more efficiently, with less waste and less creation of unwanted or unneeded illumination. Several industry groups also recognize light pollution as an important issue. For example, the Institution of Lighting Engineers in the United Kingdom provides its members information about light pollution, the problems it causes, and how to reduce its impact. Since not everyone is irritated by the same lighting sources, it is common for one person's light "pollution" to be light that is desirable for another. One example of this is found in advertising, when an advertiser wishes for particular lights to be bright and visible, even though others find them annoying. Other types of light pollution are more certain. For instance, light that *accidentally* crosses a property boundary and annoys a neighbor is generally wasted and pollutive light.

Disputes are still common when deciding appropriate action, and differences in opinion over what light is considered reasonable, and who should be responsible, mean that negotiation must sometimes take place between parties. Where objective measurement is desired, light levels can be quantified by field measurement or mathematical modeling, with results typically displayed as an isophote map or light contour map. Authorities have also taken a variety of measures for dealing with light pollution, depending on the interests, beliefs and understandings of the society involved. Measures range from doing nothing at all, to implementing strict laws and regulations about how lights may be installed and used.

Types

Light pollution is a broad term that refers to multiple problems, all of which are caused by inefficient, unappealing, or (arguably) unnecessary use of artificial light. Specific categories of light pollution include light trespass, over-illumination, glare, light clutter, and skyglow. A single offending light source often falls into more than one of these categories.

Light Trespass

Light trespass occurs when unwanted light enters one's property, for instance, by shining over a neighbor's fence. A common light trespass problem occurs when a strong light enters the window of one's home from the outside, causing problems such as sleep deprivation or the blocking of an evening view.

A number of cities in the U.S. have developed standards for outdoor lighting to protect the rights of their citizens against light trespass. To assist them, the International Dark-Sky Association has developed a set of model lighting ordinances. The Dark-Sky Association was started to reduce the light going up into the sky which reduces visibility of stars. This is any light which is emitted more than 90° above nadir. By limiting light at this 90° mark they have also reduced the light output in the 80–90° range which creates most of the light trespass issues. U.S. federal agencies may also enforce standards and process complaints within their areas of jurisdiction. For instance, in the case of light trespass by white strobe lighting from communication towers in excess of FAA minimum lighting requirements the Federal Communications Commission maintains an Antenna Structure Registration database information which citizens may use to identify offending structures and provides a mechanism for processing consumer inquiries and complaints. The US Green Building Council (USGBC) has also incorporated a credit for reducing the amount of light trespass and sky glow into their environmentally friendly building standard known as LEED.

Light trespass can be reduced by selecting light fixtures which limit the amount of light emitted more than 80° above the nadir. The IESNA definitions include full cutoff (0%), cutoff (10%), and semi-cutoff (20%). (These definitions also include limits on light emitted above 90° to reduce sky glow.)

Over-illumination

A composite image of the Earth at night in 1994–95. Over-illumination is the excessive use of light. Specifically within the United States, over-illumination is responsible for approximately two million barrels of oil per day in energy wasted. This is based upon U.S. consumption of equivalent of 50 million barrels per day (7,900,000 m^3/d) of petroleum. It is further noted in the same U.S. Department of Energy source that over 30% of all energy is consumed by commercial, industrial and residential sectors. Energy audits of existing buildings demonstrate that the lighting component of residential, commercial and industrial uses consumes about 20–40% of those land uses, variable with region and land use. (Residential use lighting consumes only 10–30% of the energy bill while commercial buildings major use is lighting.) Thus lighting energy accounts for about four or five million barrels of oil (equivalent) per day. Again energy audit data demonstrates that about 30–60% of energy consumed in lighting is unneeded or gratuitous.

An alternative calculation starts with the fact that commercial building lighting consumes in excess of 81.68 terawatts (1999 data) of electricity, according to the U.S. DOE. Thus commercial lighting alone consumes about four to five million barrels per day (equivalent) of petroleum, in line with the alternate rationale above to estimate U.S. lighting energy consumption.

Over-illumination stems from several factors:

- Not using timers, occupancy sensors or other controls to extinguish lighting when not needed;
- Improper design, especially of workplace spaces, by specifying higher levels of light than needed for a given task;
- Incorrect choice of fixtures or light bulbs, which do not direct light into areas as needed;
- Improper selection of hardware to utilize more energy than needed to accomplish the lighting task;
- Incomplete training of building managers and occupants to use lighting systems efficiently;
- Inadequate lighting maintenance resulting in increased stray light and energy costs;
- "Daylight lighting" demanded by citizens to reduce crime or by shop owners to attract customers;

- Substitution of old mercury lamps with more efficient sodium or metal halide lamps using the same electrical power; and,
- Indirect lighting techniques, such as lighting a vertical wall to bounce photons on the ground.

Most of these issues can be readily corrected with available, inexpensive technology, and with resolution of landlord/tenant practices that create barriers to rapid correction of these matters. Most importantly public awareness would need to improve for industrialized countries to realize the large payoff in reducing over-illumination.

Glare

Glare can be categorized into different types. One such classification is described in a book by Bob Mizon, coordinator for the British Astronomical Association's Campaign for Dark Skies. According to this classification:

- *Blinding glare* describes effects such as that caused by staring into the Sun. It is completely blinding and leaves temporary or permanent vision deficiencies.
- *Disability glare* describes effects such as being blinded by oncoming car lights, or light scattering in fog or in the eye, reducing contrast, as well as reflections from print and other dark areas that render them bright, with significant reduction in sight capabilities.
- *Discomfort glare* does not typically cause a dangerous situation in itself, though it is annoying and irritating at best. It can potentially cause fatigue if experienced over extended periods.

According to Mario Motta, president of the Massachusetts Medical Society, "... glare from bad lighting is a public-health hazard—especially the older you become. Glare light scattering in the eye causes loss of contrast and leads to unsafe driving conditions, much like the glare on a dirty windshield from low-angle sunlight or the high beams from an oncoming car." In essence bright and/or badly shielded lights around roads can partially blind drivers or pedestrians and contribute to accidents.

The blinding effect is caused in large part by reduced contrast due to light scattering in the eye by excessive brightness, or to reflection of light from dark areas in the field of vision, with luminance similar to the background luminance. This kind of glare is a particular instance of disability glare, called veiling glare. (This is not the same as loss of accommodation of night vision which is caused by the direct effect of the light itself on the eye.)

Light Clutter

Light clutter refers to excessive groupings of lights. Groupings of lights may generate confusion, distract from obstacles (including those that they may be intended to illuminate), and potentially cause accidents. Clutter is particularly noticeable on roads where the street lights are badly designed, or where brightly lit advertising surrounds the roadways. Depending on the motives of the person or organization that installed the lights, their placement and design can even be intended to distract drivers, and can contribute to accidents. Clutter may also present a hazard in the aviation environment if aviation safety lighting must compete for pilot attention with non-relevant lighting. For instance, runway lighting may be confused with an array of suburban commercial lighting and aircraft collision avoidance lights may be confused with ground lights.

Skyglow

Skyglow refers to the "glow" effect that can be seen over populated areas. It is the combination of all light reflected from what it has illuminated escaping up into the sky and from *all* of the badly directed light in that area that also escapes into the sky, being scattered (redirected) by the atmosphere back toward the ground. This scattering is very strongly related to the wavelength of the light when the air is very clear (with very little aerosols). Rayleigh scattering dominates in such clear air, making the sky appear blue in the daytime. When there is significant aerosol (typical of most modern polluted conditions), the scattered light has less dependence on wavelength, making a whiter daytime sky. Because of this Rayleigh effect, and because of the eye's increased sensitivity to white or blue-rich light sources when adapted to very low light levels, white or blue-rich light contributes significantly more to sky-glow than an equal amount of yellow light. Sky glow is of particular irritation to astronomers, because it reduces contrast in the night sky to the extent where it may even become impossible to see any but the brightest stars.

The Bortle Dark-Sky Scale, originally published in *Sky & Telescope* magazine, is sometimes used (by groups like the U.S. National Park Service) to quantify skyglow and general sky clarity. The nine-class scale rates the darkness of the night sky and the visibility of its phenomena, such as the gegenschein and the zodiacal light (easily masked by skyglow), providing a detailed description of each level on the scale (with Class 1 being the best). Light is particularly problematic for amateur astronomers, whose ability to observe the night sky from their property is likely to be inhibited by any stray light from nearby. Most major optical astronomical observatories are surrounded by zones of strictly enforced restrictions on light emissions.

"Direct" skyglow is reduced by selecting lighting fixtures which limit the amount of light emitted more than 90° above the nadir. The IESNA definitions include full cutoff (0%), cutoff (2.5%), and semi-cutoff (5%). "Indirect" skyglow produced by reflections from vertical and horizontal surfaces is harder to manage; the only effective method for preventing it is by minimizing over-illumination. But it has to be taken into account that according to late 2010 publications reported by Monthly Notices of Royal Astronomical Society ("Campaign of sky brightness and extinction measurements by using a portable CCD Camera", F. Falchi) Italian regions using full cut off lighting only does not increase skyglow. Anyway light reflected upwards by dark surfaces such as roads or building can be considered as minor, so debate about contribution of "indirect" skyglow will last long.

Skyglow is made considerably worse when clouds are present. While this has no effect on astronomical observations (which are not possible at visible wavelengths under cloud cover), it is very important in the context of ecological light pollution. Since cloudy nights can be up to ten times brighter than clear nights, any organisms that are affected by sky glow (e.g. zooplankton and fish that visually prey on them) are much more likely to have their ordinary behavior disturbed on cloudy nights.

Measurement and Global Effects

Measuring the effect of sky glow on a global scale is a complex procedure. The natural atmosphere is not completely dark, even in the absence of terrestrial sources of light and illumination from the Moon. This is caused by two main sources: *airglow* and *scattered light*.

At high altitudes, primarily above the mesosphere, there is enough UV radiation from the sun of very short wavelength that ionization occurs. When these ions collide with electrically neutral particles they recombine and emit photons in the process, causing airglow. The degree of ionization is sufficiently large to allow a constant emission of radiation even during the night when the upper atmosphere is in the Earth's shadow. Lower in the atmosphere all of the solar photons with energies above the ionization potential of N_2 and O_2 have already been absorbed by the higher layers and thus no appreciable ionization occurs. Apart from emitting light, the sky also scatters incoming light, primarily from distant stars and the Milky Way, but also the zodiacal light, sunlight that is reflected and backscattered from interplanetary dust particles.

The amount of airglow and zodiacal light is quite variable (depending, amongst other things on sunspot activity and the Solar cycle) but given optimal conditions the darkest possible sky has a brightness of about 22 magnitude/square arcsecond. If a full moon is present, the sky brightness increases to 18 magnitude/sq. arcsecond, 40 times brighter than the darkest sky. In densely populated areas a sky brightness of 17 magnitude/sq. arcsecond is not uncommon, or as much as 100 times brighter than is natural. To precisely measure how bright the sky gets, night time satellite imagery of the earth is used as raw input for the number and intensity of light sources. These are put into a physical model of scattering due to air molecules and aerosoles to calculate cumulative sky brightness. Maps that show the enhanced sky brightness have been prepared for the entire world.

Inspection of the area surrounding Madrid reveals that the effects of light pollution caused by a single large conglomeration can be felt up to 100 km (62 mi) away from the center. Global effects of light pollution are also made obvious. The entire area consisting of southern England, Netherlands, Belgium, west Germany, and northern France have a sky brightness of at least 2 to 4 times above normal. The only places in continental Europe where the sky can attain its natural darkness is in northern Scandinavia and in islands far from the continent. In North America the situation is comparable. From the east coast to west Texas up to the Canadian border there is very significant global light pollution.

Consequences

Energy Waste: Lighting is responsible for one-fourth of all electricity consumption worldwide, and case studies have shown that several forms of over-illumination constitute energy wastage, including non-beneficial upward direction of night-time lighting. In 2007, Terna, the company responsible for managing electricity flow in Italy, reported a saving of 645.2 million kWh in electricity consumption during the daylight saving period from April to October. It attributes this saving to the delayed need for artificial lighting during the evenings. Public lighting is the single largest source of local government's greenhouse gas emissions, typically accounting for 30 to 50% of their emissions. There are 1.94 million public lights — one for every 10 Australians — that annually cost A$210 million, use 1,035 GWh of electricity and are responsible for 1.15 million tonnes of CO2 emissions.

Current public lighting in Australia, particularly for minor roads and streets, uses large amounts of energy and financial resources, while often failing to provide high quality lighting. There are many ways to improve lighting quality while reducing energy use and greenhouse gas emissions as well as lowering costs.

Disruption of Ecosystems

When artificial light affects organisms and ecosystems it is called ecological light pollution. While light at night can be beneficial, neutral, or damaging for individual species, its presence invariably disturbs ecosystems. For example, some species of spiders avoid lit areas, while other species are happy to build their spider web directly on a lamp post. Since lamp posts attract many flying insects, the spiders that don't mind light gain an advantage over the spiders that avoid it. This is a simple example of the way in which species frequencies and food webs can be disturbed by the introduction of light at night.

Light pollution poses a serious threat in particular to nocturnal wildlife, having negative impacts on plant and animal physiology. It can confuse animal navigation, alter competitive interactions, change predator-prey relations, and cause physiological harm. The rhythm of life is orchestrated by the natural diurnal patterns of light and dark, so disruption to these patterns impacts the ecological dynamics.

Studies suggest that light pollution around lakes prevents zooplankton, such as Daphnia, from eating surface algae, helping cause algal blooms that can kill off the lakes' plants and lower water quality. Light pollution may also affect ecosystems in other ways. For example, lepidopterists and entomologists have documented that nighttime light may interfere with the ability of moths and other nocturnal insects to navigate. Night-blooming flowers that depend on moths for pollination may be affected by night lighting, as there is no replacement pollinator that would not be affected by the artificial light. This can lead to species decline of plants that are unable to reproduce, and change an area's longterm ecology. A 2009 study also suggests deleterious impacts on animals and ecosystems because of perturbation of polarized light or artificial polarisation of light (even during the day, because direction of natural polarization of sun light and its reflexion is a source of information for a lot of animals). This form of pollution is named polarized light pollution (PLP). Unnatural polarized light sources can trigger maladaptive behaviors in polarization-sensitive taxa and alter ecological interactions.

Lights on tall structures can disorient migrating birds. Estimates by the U.S. Fish and Wildlife Service of the number of birds killed after being attracted to tall towers range from 4 to 5 million per year to an order of magnitude higher. The Fatal Light Awareness Program (FLAP) works with building owners in Toronto, Canada and other cities to reduce mortality of birds by turning out lights during migration periods. Similar disorientation has also been noted for bird species migrating close to offshore production and drilling facilities. Studies carried out by Nederlandse Aardolie Maatschappij b.v. (NAM) and Shell have led to development and trial of new lighting technologies in the North Sea. In early 2007, the lights were installed on the Shell production platform L15. The experiment proved a great success since the number of birds circling the platform declined by 50 to 90%.

Sea turtle hatchlings emerging from nests on beaches are another casualty of light pollution. It is a common misconception that hatchling sea turtles are attracted to the moon. Rather, they find the ocean by moving away from the dark silhouette of dunes and their vegetation, a behavior with which artificial lights interfere. The breeding activity and reproductive phenology of toads, however, are cued by moonlight. Juvenile seabirds may also be disoriented by lights as they leave their nests and fly out to sea. Amphibians and reptiles are also affected by light pollution. Introduced light sources during normally dark periods can disrupt levels of melatonin production. Melatonin is a hormone that regulates photoperiodic physiology and behaviour. Some species of frogs and salamanders utilize a light-dependent "compass" to orient their migratory behaviour to breeding sites. Introduced light can also cause developmental irregularities, such as retinal damage, reduced sperm production, and genetic mutation.

Effect on Astronomy

Astronomy, both amateur and professional, is very sensitive to light pollution. The night sky viewed from a city bears no resemblance to what can be seen from dark skies. Skyglow (the scattering of light in the atmosphere) reduces the contrast between stars and galaxies and the sky itself, making it very much harder see fainter objects. This is one factor that has caused newer telescopes to be built in increasingly remote areas. Some astronomers use narrow-band "nebula filters" which only allow specific wavelengths of light commonly seen in nebulae, or broad-band "light pollution filters" which are designed to reduce (but not eliminate) the effects of light pollution by filtering out spectral lines commonly emitted by sodium- and mercury-vapor lamps, thus enhancing contrast and improving the view of dim objects such as galaxies and nebulae. Unfortunately these light pollution reduction (LPR) filters are not a cure for light pollution. LPR filters reduce the brightness of the object under study and this limits the use of higher magnifications. LPR filters work by blocking light of certain wavelengths, which alters the color of the object, often creating a pronounced green cast. Furthermore, LPR filters only work on certain object types (mainly emission nebulae) and are of little use on galaxies and stars. No filter can match the effectiveness of a dark sky for visual or photographic purposes. Due to their low surface brightness, the visibility of diffuse sky objects such as nebulae and galaxies is affected by light pollution more than are stars. Most such objects are rendered invisible in heavily light polluted skies around major cities. A simple method for estimating the darkness of a location is to look for the Milky Way, which from truly dark skies appears bright enough to cast a shadow.

In addition to skyglow, light trespass can impact observations when artificial light directly enters the tube of the telescope and is reflected from non-optical surfaces until it eventually reaches the eyepiece. This direct form of light pollution causes a glow across the field of view which reduces contrast. Light trespass also makes it hard for a visual observer to become sufficiently dark adapted. The usual measures to reduce this glare, if reducing the light directly is not an option, include flocking the telescope tube and accessories to reduce reflection, and putting a light shield (also usable as a dew shield) on the telescope to reduce light entering from angles other than those near the target. Under these conditions, some astronomers prefer to observe under a black cloth to ensure maximum dark adaptation. In one Italian regional lighting code this effect of stray light is defined as "optical pollution", due to the fact that there is a direct path from the light source to the "optic" – the observer's eye or telescope.

Increase in Atmospheric Pollution

A study presented at the American Geophysical Union meeting in San Francisco found that light pollution destroys nitrate radicals thus preventing the normal night time reduction of atmospheric smog produced by fumes emitted from cars and factories. The study was presented by Harald Stark from the National Oceanic and Atmospheric Administration.

Reduction

Reducing light pollution implies many things, such as reducing sky glow, reducing glare, reducing light trespass, and reducing clutter. The method for best reducing light pollution, therefore, depends on exactly what the problem is in any given instance. Possible solutions include:

- Utilizing light sources of minimum intensity necessary to accomplish the light's purpose.
- Turning lights off using a timer or occupancy sensor or manually when not needed.
- Improving lighting fixtures, so that they direct their light more accurately towards where it is needed, and with less side effects.
- Adjusting the *type* of lights used, so that the light waves emitted are those that are less likely to cause severe light pollution problems. Mercury, metal halide and above all first generation of blue-light LED road luminaries are much more pollutant that sodium lamps: Earth atmosphere scatters and transmits blue light better than yellow or red light. It is a common experience observing "glare" and "fog" around and below LED road luminaries as soon as air humidity increases, while orange sodium lamp luminaries are less prone to show this phenomenon.
- Evaluating existing lighting plans, and re-designing some or all of the plans depending on whether existing light is actually needed.

Improving Lighting Fixtures

The use of *full cutoff* lighting fixtures, as much as possible, is advocated by most campaigners for the reduction of light pollution. It is also commonly recommended that lights be spaced appropriately for maximum efficiency, and that lamps within the fixtures not be overpowered. Full cutoff fixtures first became available in 1959 with the introduction of General Electric's M100 fixture.

A full cutoff fixture, when correctly installed, reduces the chance for light to escape above the plane of the horizontal. Light released above the horizontal may sometimes be lighting an intended target, but often serves no purpose. When it enters into the atmosphere, light contributes to sky glow. Some governments and organizations are now considering, or have already implemented, full cutoff fixtures in street lamps and stadium lighting. The use of full cutoff fixtures help to reduce sky glow by preventing light from escaping above the horizontal. Full cutoff typically reduces the visibility of the lamp and reflector within a luminaire, so the effects of glare are also reduced. Campaigners also commonly argue that full cutoff fixtures are more efficient than other fixtures, since light that would otherwise have escaped into the atmosphere may instead be directed towards the ground. However, full cutoff fixtures may also trap more light in the fixture than other types of luminaires, corresponding to lower luminaire efficiency, suggesting a re-design of some luminaires may be necessary.

The use of full cutoff fixtures can allow for lower wattage lamps to be used in the fixtures, producing the same or sometimes a better effect, due to being more carefully controlled. In every lighting system, some sky glow also results from light reflected from the ground. This reflection can be reduced, however, by being careful to use only the lowest wattage necessary for the lamp, and setting spacing between lights appropriately. Assuring luminaire setback is greater than 90° from highly reflective surfaces also diminishes reflectance. A common criticism of full cutoff lighting fixtures is that they are sometimes not as aesthetically pleasing to look at. This is most likely because historically there has not been a large market specifically for full cutoff fixtures, and because people typically like to see the source of illumination. Due to the specificity with their direction of light, full cutoff fixtures sometimes also require expertise to install for maximum effect.

The effectiveness of using full cutoff roadway lights to combat light pollution has also been called into question. According to design investigations, luminaires with full cutoff distributions (as opposed to *cutoff* or *semi cutoff*, compared here) have to be closer together to meet the same light level, uniformity and glare requirements specified by the IESNA. These simulations optimized the height and spacing of the lights while constraining the overall design to meet the IESNA requirements, and then compared total uplight and energy consumption of different luminaire designs and powers. Cutoff designs performed better than full cutoff designs, and semi-cutoff performed better than either cutoff or full cutoff. This indicates that, in roadway installations, over-illumination or poor uniformity produced by full cutoff fixtures may be more detrimental than direct uplight created by fewer cutoff or semi-cutoff fixtures. Therefore, the overall performance of existing systems could be improved more by reducing the number of luminaires than by switching to full cutoff designs.

However, using the definition of "light pollution" from some Italian regional bills (i.e., "every irradiance of artificial light outside competence areas and particularly upward the sky") only full cutoff design prevents light pollution. The Italian Lombardy region, where only full cutoff design is allowed (Lombardy act no. 17/2000, promoted by Cielobuio-coordination for the protection of the night sky), in 2007 had the lowest per capita energy consumption for public lighting in Italy: this information can be verified using data released by Terna company. The same legislation also imposes a minimum distance between street lamps of about four times their height, so full cut off street lamps are the best solution to reduce both light pollution and electrical power usage.

Adjusting Types of Light Sources

Several different types of light sources exist, each having different properties that affect their appropriateness for certain tasks, particularly efficiency and spectral power distribution. It is often the case that inappropriate light sources have been selected for a task, either due to ignorance or because more sophisticated light sources were unavailable at the time of installation. Therefore, badly chosen light sources often contribute unnecessarily to light pollution and energy waste. By re-assessing and changing the light sources used, it is often possible to reduce energy use and pollutive effects while simultaneously greatly improving efficiency and visibility.

Some types of light sources are listed in order of energy efficiency in the table below.

Type of light source	*Color*	*Luminous effectiveness (in lumens per watt)*
Low Pressure Sodium (LPS/SOX)	yellow/amber	80–200
High Pressure Sodium (HPS/SON)	pink/amber-white	90–130
Metal Halide	bluish-white/white	60–120
Mercury-Vapour	blue-greenish white	13–48
Incandescent	yellow/white	8–25

Many astronomers request that nearby communities use low pressure sodium lights as much as possible, because the principal wavelength emitted is comparably easy to work around or in rare cases filter out. The low cost of operating sodium lights is another feature. In 1980, for example, San Jose, California, replaced all street lamps with low pressure sodium lamps, whose light is easier for nearby Lick Observatory to filter out. Similar programs are now in place in Arizona and Hawaii.

Disadvantages of low pressure sodium lighting are that fixtures must usually be larger than competing fixtures, and that color cannot be distinguished, due to its emitting principally a single wavelength of light. Due to the substantial size of the lamp, particularly in higher wattages such as 135 W and 180 W, control of light emissions from low pressure sodium luminaires is more difficult. For applications requiring more precise direction of light (such as narrow roadways) the native lamp efficacy advantage of this lamp type is decreased and may be entirely lost compared to high pressure sodium lamps. Allegations that this also leads to higher amounts of light pollution from luminaires running these lamps arise principally because of older luminaires with poor shielding, still widely in use in the UK and in some other locations. Modern low-pressure sodium fixtures with better optics and full shielding, and the decreased skyglow impacts of yellow light preserve the luminous efficacy advantage of low-pressure sodium and result in most cases is less energy consumption and less visible light pollution. Unfortunately, due to continued lack of accurate information, many lighting professionals continue to disparage low-pressure sodium, contributing to its decreased acceptance and specification in lighting standards and therefore its use. Another disadvantage of low-pressure sodium lamps is that some people find the characteristic yellow light very displeasing aesthetically.

Re-designing Lighting Plans

In some cases, evaluation of existing plans has determined that more efficient lighting plans are possible. For instance, light pollution can be reduced by turning off unneeded outdoor lights, and only lighting stadiums when there are people inside. Timers are especially valuable for this purpose. One of the world's first coordinated *legislative* efforts to reduce the adverse effect of this pollution on the environment began in Flagstaff, Arizona, in the U.S. There, over three decades of ordinance development has taken place, with the full support of the population, often with government support, with community advocates, and with the help of major local observatories, including the United States Naval Observatory Flagstaff Station. Each component helps to educate, protect and enforce the imperatives to intelligently reduce detrimental light pollution.

One example of a lighting plan assessment can be seen in a report originally commissioned by the Office of the Deputy Prime Minister in the United Kingdom, and now available through the Department for Communities and Local Government. The report details a plan to be implemented throughout the UK, for designing lighting schemes in the countryside, with a particular focus on preserving the environment. In another example, the city of Calgary has recently replaced most residential street lights with models that are comparably energy efficient. The motivation is primarily operation cost and environmental conservation. The costs of installation are expected to be regained through energy savings within six to seven years. A newer method for predicting and measuring various aspects of light pollution was described in the journal Lighting Research Technology (September 2008). Scientists at Rensselaer Polytechnic Institute's Lighting Research Center have developed a comprehensive method called Outdoor Site-Lighting Performance (OSP), which allows users to quantify, and thus optimize, the performance of existing and planned lighting designs and applications to minimize excessive or obtrusive light leaving the boundaries of a property. OSP can be used by lighting engineers immediately, particularly for the investigation of glow and trespass (glare analyses are more complex to perform and current commercial software does not readily allow them), and can help users compare several lighting design alternatives for the same site.

In the effort to reduce light pollution, researchers have developed a "Unified System of Photometry," which is a way to measure how much or what kind of street lighting is needed. The Unified System of Photometry allows light fixtures to be designed to reduce energy use while maintaining or improving perceptions of visibility, safety, and security. There was a need to create a new system of light measurement at night because the biological way in which the eye's rods and cones process light is different in nighttime conditions versus daytime conditions. Using this new system of photometry, results from recent studies have indicated that replacing traditional, yellowish, high-pressure sodium (HPS) lights with "cool" white light sources, such as induction, fluorescent, ceramic metal halide, or LEDs can actually reduce the amount of electric power used for lighting while maintaining or improving visibility in nighttime conditions.

On the other hand it has to be remembered that human night vision in better in blue or white light only if light intensity is low, while lighting and power industries are pushing to increase more and more ground illuminances: to avoid a conflict of interests ground illuminance or luminances required by standards has to be dropped down if blue or white light sources are used. Moreover blue and white light scatters more than yellow or red light, so overall light pollution could be increased by metal halide and above all by LED street lamps while fluorescent lamps usually include small quantities of mercury and should be treated as special waste. The International Commission on Illumination, also known as the CIE from its French title, le Commission Internationale de l'Eclairage, will soon be releasing its own form of unified photometry for outdoor lighting.

Bibliography

Asaithambi, S.: *Economics of Ground Water Management in India*, Abhijeet, Delhi, 2008.

Bauer, Steffan: *Handbook of Globalization and the Environment*, CRC Press, Delhi, 2007.

Berkoff, Jeremy: *A Strategy for Managing Water in the Middle East and North Africa*, World Bank, Washington, DC, 1994.

Bhave, P.R. and R. Gupta: *Analysis of Water Distribution Networks*, Narosa, Delhi, 2011.

Bucks D.A., Nakayama F.S. and Gilbert R.G.: *Trickle Irrigation Water Quality and Preventive Maintenance*, Agricultural Water Management, NY, 1979.

Collier, W; Webb, R.: *Floods, Droughts and Climate Change*, University of Arizona Press, NY, 2001.

Conway, D.: *Climate Change and Water Resources in the Nile Basin*, University of London, London, 1993.

Duda Alfred M.: *"Environment Considerations in Implementing the Comprehensive Approach to Water Management"*, Washington, 1993.

Dyson, Tim: *India's Historical Demography: Studies in Famine, Disease, and Society*, Riverdale, Westwood, MA, 1989.

Elmusa, Sharif S.: *Water Conflict: Economics, Politics, Law and Palestinian-Israeli Water Resources*, Institute for Palestine Studies, Beirut, Lebanon, 1998.

Ferentinos L.: *Proceeding of the Sustainable Taro Culture for the Pacific Conference*, Honolulu, HITAHR, 1993.

Gopalan Swapna : *Natural Disaster Insurance*, Icfai University Press, Delhi, 2008.

Gupta G.D.: *Management of Natural Disasters in Developing Countries* , Daya, Delhi, 2006.

Gupta K.R.: *Water Crisis in India*, Atlantic Pub, Delhi, 2008.

Hillel, Daniel: *Rivers of Eden: the Struggle for Water and the Quest for Peace in the Middle East*, Oxford University Press, NYC, 1994.

Jana B. L.: *Water Harvesting and Watershed Management*, Agrotech, Delhi, 2008.

Kumar, Arvind : *Environment and Global Warming*, Shree Pub, Delhi, 2009.

Lewis, Peirce. *New Orleans: The Making of an Urban Landscape*. Cambridge, MA: Ballinger Publishing Co., 1976. Annotation

Metcalf and Eddy Inc: *Waste-water Engineering: Treatment, Disposal, Reuse*, McGraw-Hill Book Co, New York, 1979.

Monroe, D.W. and Key, W. P.: *The Feasibility of Ozone for Purification of Hatchery Waters*;Ozone Science and Engineering, 1980.

Nelson, Carl L.: *Protecting the Past from Natural Disasters*, The Preservation Press, National Trust for Historic Preservation, 1991.

Newson, M. D.: *Flood Effectiveness in River Basins: Progress in Britain in a Decade of Drought*, John Wiley and Sons, Chichester, 1989.

Pacey A., and A. Cullis: *Rainwater Harvesting: The Collection of Rainfall and Runoff in Rural Areas*, London, U.K.: IT Publications, 1986.

Paul A. Keddy: *Wetland Ecology – Principles and Conservation*, Cambridge University Press, Cambridge, 2000.

Reiter, L.: *Earthquake Hazard Analysis: Issues and Insights*, Columbia University Press, New York, 1990.

Ritchie, J.: *Landscape and Material Life in Franklin County, Massachusetts, 1770-1860*. Knoxville: University of Tennessee Press, 1991.

Shamsi, U.M.: *GIS Applications for Water, Wastewater, and Stormwater Systems*, Routledge, Delhi, 2010.

Singh A.K.: *Environment and Water Resources Management*, Adhyayan, Delhi, 2006.

Swarnim Kumari : *Nuclear Weapons, Global Warming and Disaster Management*, Rajat Pub, Delhi, 2010.

Thomas, Harold E.: *The Conservation of Groundwater: A Survey of the Present Groundwater Situation in the U.S.*, McGraw Hill, New York, NY. 1951.

William Graebner: *Coal Mining Safety in the Progressive Era: The Political Economy of Reform*, U of Kentucky Press, NY, 1976.

Winfried Lampert and Ulrich Sommer: *Limnoecology – The Ecology of Lakes and Streams*, Oxford University Press, NY, 1997.

Ziminska, H.: *Protein Recovery from Fish Wastewaters*; American Society of Agricultural Engineering; MI, U.S.A., 1985.

Index

❑❑❑